OECD Review of Risk Management Policies Morocco

OECD
BETTER POLICIES FOR BETTER LIVES

This work is published under the responsibility of the Secretary-General of the OECD. The opinions expressed and arguments employed herein do not necessarily reflect the official views of OECD member countries.

This document and any map included herein are without prejudice to the status of or sovereignty over any territory, to the delimitation of international frontiers and boundaries and to the name of any territory, city or area.

Please cite this publication as:
OECD (2017), *OECD Review of Risk Management Policies Morocco*, OECD Publishing, Paris.
http://dx.doi.org/10.1787/9789264276482-en

ISBN 978-92-64-27638-3 (print)
ISBN 978-92-64-27648-2 (PDF)

The statistical data for Israel are supplied by and under the responsibility of the relevant Israeli authorities. The use of such data by the OECD is without prejudice to the status of the Golan Heights, East Jerusalem and Israeli settlements in the West Bank under the terms of international law.

Photo credits: Cover © iStockophoto.com/Pavliha; © wayfarerlife; © Maghreb Arabe Presse (MAP).

Corrigenda to OECD publications may be found on line at: *www.oecd.org/about/publishing/corrigenda.htm*.

Foreword

A fundamental responsibility of public authorities, but also of all social and business stakeholders, is to ensure resilience to critical risks. Such resilience is needed to guarantee people's safety and well-being, enable sustainable economic growth, and maintain confidence in public and economic institutions. Morocco is exposed to a number of risks that have an increasing probability of occurring. These risks are weather-related in particular, but others are of geological, technological or human origin. The sources of the country's vulnerabilities have increased in tandem with economic development, against a backdrop of rapid urbanisation, a concentration of certain activities on the coast, and a growing exposure to the effects of climate change.

Morocco undertook serious efforts over the past decade to boost its resilience to critical risks and these have created the conditions for the country to adopt an enhanced framework for risk management. Specifically, the political commitment demonstrated after recent fatal flooding, the changes expected from the ongoing decentralisation process, and citizens' increased expectations for improved governance offer a major opportunity that needs to be seized. The organisation of the 22nd session of the Conference of the Parties to the United Nations Framework Convention on Climate Change, held in November 2016 in Marrakesh, also helped focus attention on this issue.

The *OECD Review of Risk Management Policies in Morocco* is based on a self-assessment by national and regional stakeholders, renowned international experts, and using the OECD's peer review methodology. The report identifies both the successes and strengths of these policies, as well as the related challenges that still need to be addressed. It makes concrete recommendations for improving critical risk management, using a holistic, inclusive and integrated approach. Through these recommendations, public authorities in Morocco are encouraged to build a strategic long-term vision to ensure better risk management, by strengthening assessment, prevention and emergency and recovery policies, while making this issue a public policy priority. Going forward, it is the responsibility of national and local authorities, as well as other stakeholders, to consider and implement these recommendations. The ultimate goal is to enhance the operation of Morocco's current risk management systems and the resilience of its economy and society, when confronted with critical risks.

This study was undertaken as part of the OECD High Level Risk Forum, organised by the Directorate for Public Governance and Territorial Development. The analytical framework is based on the OECD Recommendation on the Governance of Critical Risks, adopted by the OECD Council on 6 May 2014. This Recommendation contributed to the Sendai Framework for Disaster Risk Reduction 2015-2030, which was adopted by the UN in March 2015. This review is also part of a series of OECD Reviews of risk management policies, including similar studies on France, Italy, Japan and Mexico.

This review is part of a broader programme of building closer relations between the OECD and Morocco, as emphasised by the signature of a "Country Programme" cooperation agreement signed on 15 June 2015 between the OECD and the Kingdom of Morocco.

The OECD supports Morocco's efforts to improve resilience in the future.

ACKNOWLEDGEMENTS

The *OECD Review of Risk Management Policies in Morocco* was carried out by the Directorate for Public Governance and Territorial Development (GOV), under the responsibility of Rolf Alter, and as part of the OECD High Level Risk Forum. The project was overseen by Charles Baubion, under the supervision of Stéphane Jacobzone, Deputy Head of the Reform of the Public Sector Division. Jack Radisch, Senior Project Manager of the High Level Risk Forum, provided general guidance throughout the process. This report was written by Charles Baubion, Aliae Sayah and Pierre-Alain Schieb. Meryem Guelzim provided support with data and information collection. The authors would like to thank Luiz de Mello, Deputy Director of the Directorate for Public Governance and Territorial Development, for his strategic notes.

The OECD Secretariat benefitted from the expertise and comments of three international experts (peers) on risk management: Walter Amman (Global Risk Forum, Davos, Switzerland), Orsola Lussignoli (Federal Office of Civil Protection and Disaster Assistance, Germany) and Régis Thépot (Seine Grands Lacs River Basin Authority, France). Erika Conti (Civil Protection Department, Italy) also took part in the national political dialogue.

The Secretariat would like to extend its warm thanks to teams from the Moroccan Ministry of General Affairs and Governance and the Ministry of the Interior, who engaged all of the stakeholders involved and facilitated the information-gathering and fact-checking process, and contributed to the organisation of interviews with stakeholders as well as to the completion of the results of the review. Particular thanks go to Aziz Ajbilou, Secretary-General, Mohamed Mahdad, Director of Governance, and Azzdine El Hammoudi, Head of the Risk Management and Monitoring Department of the Ministry of General Affairs and Governance, as well as Noureddine Boutayeb, *Wali*, Secretary General, Said Ziane, Special Assistant at the Secretariat-General, and Rachid Afirat, Governor, Ministry of the Interior.

The OECD Secretariat would also like to thank all stakeholders in Morocco for their availability and engagement (see Appendix B). Their participation in the launch event on 10 April 2015 in Rabat, at the series of interviews held in May and June 2015, and in the national political dialogue on 13 October 2015 in Rabat, as well as their questionnaire answers, helped to gather the information and data that were essential to the exhaustive analysis presented in this report.

Special thanks should also go to Olaf Handloegten (Global Initiative on Disaster Risk Management) of GIZ, the German Corporation for International Cooperation working on behalf of the German Federal Ministry of Economic Cooperation and Development, which offered its support to the project. The project also benefitted from a contribution of the Moroccan Ministry of General Affairs and Governance.

Editorial and communication services were provided by Lynda Hawe, Maha Skah, Andrea Uhrhammer, Liz Zachary.

Table of contents

Figures

Tables

Boxes

List of acronyms

ABH	Hydraulic Basin Agency
AFD	French Development Agency
AfDB	African Development Bank
ANP	National Port Authority
ANPME	National Agency for the Promotion of SMEs
ASVTS	Life and Earth Sciences Association
BCP	Business Continuity Plan
CCG	Central Guarantee Fund
CDP	Communal Development Plan
CGEM	General Confederation of Moroccan Companies
CNRST	National Centre of Scientific and Technical Research
CRTS	Royal Centre for Remote Sensing
CVC	Monitoring and Coordination Centre
DAPS	Directorate of Insurance and Social Welfare
DDC	Swiss Agency for Development and Cooperation
DGPC	Directorate-General of Civil Protection
DMN	National Meteorological Office
FLCN	Fund to Combat the Effects of Natural Disasters
FSHIU	Housing Solidarity and Urban Integration Fund
GDP	Gross Domestic Product
GIS	Geographic Information System
HCEFLCD	High Commission for Waterways and Forests and Fight against Desertification
HCP	High Commission for Planning
IFRC	International Federation of Red Cross and Red Crescent Societies
INDH	National Human Development Initiative
ING	National Geophysical Institute
INRA	National Institute of Agricultural Research
INSARAG	International Search and Rescue Advisory Group
IPCC	Intergovernmental Panel on Climate Change
IRES	Royal Institute of Strategic Studies
LYDEC	Lyonnaise des Eaux de Casablanca
MAGG	Ministry of General Affairs and Governance
MAMDA	Mutuelle Agricole Marocaine des Assurances
MAPM	Ministry of Agriculture and Fisheries

METL	Ministry of Public Works, Transportation and Logistics
MnhPRA	Morocco Natural Hazards Probabilistic Risk Analysis
MP	Management Plan
NGO	Non-governmental organisation
OCP	Office Chérifien des Phosphates (Moroccan phosphate rock producer)
OECD	Organisation for Economic Cooperation and Development
ONCF	National Railways Authority
ONDA	National Airports Authority
ONEE	National Water and Electricity Authority
OP	Operational Plan
ORSEC	Organisation des Secours
PAN-LCD	National Plan of Action for Desertification Control
PDAIRE	Integrated Water Resources Development Plan
PNEEI	National Irrigation Water Saving Programme
PNPI	National Flood Protection Plan
SAMU	Moroccan Emergency Medical Services
SCR	Société Centrale de Réassurance
SDACR	Master Plan for Risk Analysis and Coverage
SDAU	Master Plan for Urban Development
SEFER	Railway Contingency Plan
SINON	Emergency Flooding Plan
SNAT	National Territorial Development Plan
SNE	National Water Strategy
SPAC	Flood Prevention and Warning System
UNDP	United Nations Development Programme
UNESCO	United Nations Educational, Scientific and Cultural Organisation
UNISDR	United Nations Office for Disaster Risk Reduction
USAID	United States Agency for International Development

Executive summary

The implementation of an ambitious risk management policy in Morocco is essential to guarantee continued inclusive growth and maintain public trust by combining economic progress and greater social inclusion. Morocco is exposed to a number of risks. Natural hazards as well as industrial accidents or human risks can have serious potential socio-economic consequences. The country's heterogeneous risk profile is characterised by floods that occur on an almost annual basis, increasingly frequent periods of drought, earthquake zones in the north and south of the country, and a coastline exposed to tsunamis. The significant acceleration of economic growth over the past few years has been accompanied by an increase in vulnerabilities, including urban concentration, rural exodus and coastal development, integration into the global value chains, tighter budgetary constraints, and climate change. In addition, the current social and political context is characterised by high expectations from Moroccan citizens, which is making risk management a particularly important public policy issue in the country.

In the last decade, Morocco has undertaken considerable efforts to enhance its resilience, as demonstrated by the country's adoption of the OECD Recommendation on the Governance of Critical Risks in May 2015. At the national level, regulatory and institutional reforms, as well as the establishment of new financial mechanisms, resulted in a better inclusion of risk-related issues in the formulation of public policies. A number of pilot initiatives, run by local authorities, civil society and the private sector, which benefited from the support of the international co-operation, also contribute to a constantly improving risk management system. The OECD report identified a wide range of risk management practices in Morocco that meet with high international standards in the field.

However, governance gaps still remain. This results in policies that are often too focused on a single risk, too sector-based, or not sufficiently cross-cutting or inclusive. Therefore, fully benefiting from these efforts would require Morocco to adopt a clear strategic framework for a holistic approach to risk management. This framework would need to incorporate elements of risk prevention as well as emergency response and recovery, and ensure that they become a public policy priority. This would enable key challenges to be addressed, such as the coordination among different ministerial departments and levels of government, as well as the increased inclusion of civil society, businesses and the scientific community when defining and implementing risk-related policies.

The convergence of a number of opportunities enables the establishment of an ambitious risk governance framework in Morocco. The implementation of this new framework should benefit from the strong political momentum following the floods of November/December 2014, the ongoing decentralisation process and its implications for regional authorities, the development of critical infrastructure projects, citizens' high expectations concerning the strengthening of public governance, and the organisation of the 22nd Conference of the Parties on Climate Change in November 2016. By investing more in boosting resilience, Morocco will enjoy high returns in the future, in terms of well-being, attractiveness, competitiveness and stability.

In particular, the report makes the following recommendations to public authorities in Morocco for improving the governance of critical risk:

1. Develop a holistic strategy including a common long-term vision and shared objectives, thereby giving risk-related policies the status of national priority. This strategy will need to be supported by strong institutional leadership, the establishment of multi-stakeholders consultation forums, a modernised regulatory framework, and a monitoring and evaluation process. Connecting this strategic framework with the other sectoral strategies of the country will be essential to its successful implementation.

2. Establish a national risk assessment programme that will serve as reference for all risk-related public policies. The programme should be based on a clearly defined structure, guaranteeing shared and harmonised assessment methods among all risk-related stakeholders. It must pull together and capitalise on existing university and technical expertise.

3. Strengthen risk culture among public decision makers, citizens and businesses. A risk culture should be based on operational communication and risk awareness tools.

4. Develop a framework that creates incentives for increased risk prevention and resilience efforts. The framework should allow for a better balance of priorities between structural and non-structural measures in order to optimise investments and the use of public resources. It should also be coupled with sustainable financing and transparent decision-making tools.

5. Continue to promote greater flexibility in emergency response by improving the level of preparedness of all stakeholders involved. This will be achieved by enhanced emergency planning, the widespread introduction of multi-hazard early warning systems, improved coordination between ministries and departments through inter-agency emergency plans, and the implementation of modernised crisis communication tools. Public authorities will need to strengthen their ability to scale-up and accelerate their emergency response to crises, especially by harmonising their capabilities at the national level on the basis of a shared risk assessment. They will also need to reinforce volunteer organisations, streamline international co-operation arrangements for emergencies and make post crisis feedback and learning more systematic.

6. Adopt a budgetary approach that sustains disaster risk financing in public accounts by incorporating potential future damage estimates. Subsequently, it will be necessary to combine risk transfer mechanisms, particularly in the agricultural sector, rules for clear ex-post public compensation schemes, and a transparent reconstruction process building process that will lead to enhanced resilience.

Assessment and recommendations

Significant exposure to critical risks

This study focuses on Morocco's most significant risks in terms of potential human and economic impact. These are flooding, drought, earthquakes and tsunamis, all of which affect numerous areas across the country.

- Floods cause significant impacts, including high numbers of human casualties (47 deaths in December 2014 in Guelmim Region, around 100 deaths during the Ourika floods of 1995), significant economic damage in the country's large urban centres (Tangiers, Casablanca, Agadir), a disruption of economic activity and damage to infrastructure (1bn Moroccan Dirhams' worth of damage during the Gharb floods of 2009, according to the Ministry of Agriculture).
- Drought regularly affects the country, causing major losses of agricultural GDP, which contributes up to 15% of national GDP, and severely affecting rural populations – who often have no other choice but to leave for outlying districts of large cities.
- Earthquakes, although rare, affect two specific areas of the country: the north, which is currently experiencing strong economic growth, and Agadir Region, one of Morocco's main touristic centres. In 2004, the last major earthquake in the north claimed 600 victims and destroyed 12,000 homes in Al Hoceima Region.
- The risk of tsunamis has a much lower probability. Nevertheless, such disaster could have dramatic consequences for a large, densely populated and economically important area on the Moroccan coast.

Other risks (e.g. industrial, technological, or locust outbreaks) must also be considered in a national multi-hazard approach that would benefit from the sharing and pooling of methods and resources to tackle the different risks, and thus increase the effectiveness of public policies.

Accelerated growth is accompanied by increased vulnerability...

A high growth rate has been propelled, by increased domestic demand and public investment (nearly 5% increase was recorded during the 2000-2009 period). On the one hand, this growth has improved Moroccans' well-being, especially through improved access to basic infrastructure (water, roads and electricity). On the other hand, the current international economic backdrop has increased pressures on Morocco's economic model. Indeed, its exposure to a series of shocks from 2008 onwards has contributed to increase the country's deficit and public debt.

A greater exposure to critical risks and budgetary constraints will probably force the country to take decisions in favour of a more structured risk management policy that is also more preventive and inclusive.

According to the World Bank (2013), Morocco has a 90% chance of experiencing a disaster event causing losses of 10bn Dirhams within the next 30 years. This figure represents 1% of the country's GDP in any given year. Such a shock would potentially have a negative impact on Morocco's growth path and public finances, which have been weakened by the current economic situation. Failing to correctly anticipate such risk or to introduce measures to prevent it from occurring, could result in destabilising effects. The challenge for a successful risk management policy is, thus, to create the necessary conditions for increased resilience and to facilitate the smoothing of the economic and financial consequences of a major shock over time.

… which has become a key issue for public policy and governance

An effective risk management policy contributes towards maintaining confidence in institutions and building resilient development, in line with the Sustainable Development Goals (SDGs) and the OECD Recommendation on the Governance of Critical Risks, adopted by the OECD Council on 6 May 2014. In fact, this is an area of public policy where citizens' expectations are particularly high. While questions of inclusion, transparency and accountability have been at the heart of societal demands in Morocco in recent years, risk management constitutes a decisive issue.

The significant progress made by Morocco in economic, social, environmental and cultural fields represents an opportunity to address questions of critical risk in a more assertive manner. Human capital, financial and material resources are more readily available. However, the country is undergoing rapid transformations that are characteristic of fast-developing and emerging economies, such as urbanisation and littoralisation, and the increased geographical concentration of added economic value. If dominant trends, especially demographic ones continue, vulnerability to critical risks and the impact of adverse events will increase substantially. The issue of critical risks should therefore become a strategic priority in planning and governance efforts.

An ambitious critical risk policy: a "win-win" solution?

Incorporating risk management into public policy decisions creates opportunities:

- Risk resilience is featured in international development, well-being and competitiveness indicators.
- The country's attractiveness as appreciated by citizens' and investors' partly depends on good risk management.
- The resilience of critical infrastructures and business sectors is contingent upon the quality of prevention and anticipation measures.
- Building risk management expertise can also become an exportable service (with Morocco as a regional service hub).
- A shared risk culture contributes towards societal cohesion and confidence.
- Investing in resilience has high returns: in the long term, the reduction of harmful consequences, a more equal share of risk between stakeholders, a better integration of technology, and international co-operation contribute to more sustainable economic development.

Recommendations

PRIORITISE RISK MANAGEMENT IN MOROCCO TO ACHIEVE SUSTAINABLE GROWTH IN THE SHORT, MEDIUM AND LONG TERM

The OECD Recommendation on the Governance of Critical Risks recognises that critical risks can manifest themselves quickly and unexpectedly. Citizens, meanwhile, have high expectations when it comes to governments' level of preparedness. In fact, the effective management of critical risks is a prerequisite to ensure the population's well-being and the country's competitiveness.

A more holistic approach to risk in Morocco would benefit from a favourable context at this unique moment in Moroccan history. The strong political will, the pursuit of economic growth, the development of major infrastructures, the ongoing decentralisation process and regulatory reforms, the strengthening of sector-based plans, as well as the organisation of the COP 22 meeting in 2016 provide unprecedented opportunities to pursue efforts to incorporate the "risk" dimension into decisions made by public authorities, the private sector and citizens.

Governance of critical risks: Towards a national strategy that will unify different efforts

Over the past 20 years, Morocco has made considerable progress in understanding critical risks…

Progress in institutional terms includes a legislative and regulatory body, the implementation of specific financial mechanisms, and the allocation of responsibilities to various bodies and sector-based economic development plans that incorporate risk. This has gone hand-in-hand with numerous pilot projects that were spearheaded by the private sector, civil society and decentralised public authorities. During the OECD Peer Review in Morocco, nearly 45 examples of best practices were identified.

These efforst towards improved risk management policies in Morocco benefited from Morocco's increased integration in global value chains and from support of international co-operation. However, it stemmed mainly from the country's need to respond to a series of major adverse events. The response element, which manifests itself through the importance of emergency management, has consequently become the driving force behind this increased focus on risk management.

…but these efforts have not yet led to the adoption of an integrated response…

Striking deficiencies in governance still remain, such as policies that are too sector-based, focused on a single risk, not sufficiently inclusive, using a top-down approach, or that favour an emergency response and structural approaches to risk prevention. The danger is not just general ineffectiveness and a scattered distribution of resources, but also that efforts made will not be successful. An inappropriately executed or poorly understood allocation of responsibilities and resources to stakeholders prevents the emergence of a common strategy.

A gap analysis can identify vulnerabilities in the present management system with respect to the challenges involved in putting together a comprehensive risk management approach in the short, medium and long term, such as:

- A planning/anticipation timeframe that remains insufficient.

- Governance is not sufficiently cross-disciplinary, and lacks shared, clearly defined allocation of responsibilities.
- Sub-optimal multi-stakeholder co-ordination.
- Vertical integration with the regions, provinces and municipalities that require strengthening.
- Civil society organisations need to be given stakeholder status.
- A better allocation of resources between sectors, within the risk management cycle, and over different periods of time.

...or a sufficiently inclusive risk management policy

Various stakeholders have struggled to agree on the general goals of risk management. Consequently, the co-ordination and alignment of different initiatives with a common strategy remain insufficient. Despite the involvement of a large number of key stakeholders, criteria for evaluating the performance and consistency of their respective contributions have not been defined. This performance evaluation deficiency increases the difficulty of allocating risk management-related responsibilities and resources in an optimal manner to the various initiatives. There has been, however, a continuous improvement in the consultation process and the consistency of the legislative framework at the national level.

It is time for a forward-looking approach to risk governance

Since Morocco has now reached a relatively high level of critical mass in the field of risk management policies, it is now an opportune moment to capitalise on the current situation and prepare for the governance of the future. Because many risk-related public policy issues have a long-term dimension (urbanisation, water infrastructure, transport, energy, demographics, education and risk culture), the timeframe for risk assessments must go beyond the ones of usual sector-based plans, up to 50 or even 100 years. The impact of climate change or the occurrence of tsunamis can in fact only be assessed over a much longer timeframe.

Aligning risk governance with current challenges and long timeframes requires the full mobilisation of available capacities. Stakeholders consulted in Morocco and OECD expert groups and peer reviewers strongly expressed that Morocco is now ready to deliver on high ambition levels and consequently build a governance framework at the service of an integrated risk management strategy combined with a long-term vision

Recommendations

BUILD A LONG-TERM VISION AT THE SERVICE OF AN INTEGRATED CRITICAL RISK MANAGEMENT STRATEGY

- A long-term vision that is consistent with major sector-based strategies by 2030 will set out overarching principles for risk management in Morocco.
- Provide risk policy with a status of national priority by assigning a dedicated place in the institutional organisation chart with sufficient leadership.
- Design an institutional framework with clear lines of responsibility that covers the entire risk management cycle.
- Modernise the regulatory framework and make it adaptable to current and future challenges.
- Coordinate the implementation of risk management policies with national sector-based development strategies by creating, where appropriate, links between regional and local levels.
- Establish and organise a forum to discuss and debate the general lines of action of risk management policies, fostering engagement among citizens and civil society.

Assessment: The foundation of a shared risk management policy

Risk assessment has developed significantly since the start of the 2000s…

Morocco has implemented single-risk assessment approaches, such as the Ministry of Water's efforts to assess flooding in 2003 and the seismic database launched by the Rabat Scientific Institute. These methods gradually developed into multi-risk approaches. The Ministry for the Environment published the first national risk atlas in 2008. This initiative was accompanied by the creation of a Geographic Information System gathering all of the data previously collected.

A large amount of scientific and technical expertise exists, but it remains fragmented. Tools (such as the risk atlas) and expert bodies (universities, institutes and research centres) need to be professionalised. This would involve ensuring that their funding can be sustained, their methodologies are standardised wherever possible, their gradation/scale and number of indicators become more detailed, and their databases are correctly archived and regularly updated.

Finally, a more advanced modelling attempt was recently carried out by the Moroccan Ministry for General Affairs and Governance with the support of the World Bank, and more recently the Swiss Cooperation. It enabled the creation of an integrated geographical risk assessment and modelling tool known as MnhPRA. However, its influence on practices appeared to be limited due to insufficient appropriation, communication, and training.

…But it does not allow for a comparison of different risk profiles to make trade-off decision for resource allocation…

The issues of methodology and time-horizon are crucial because various natural and associated risks have very different profiles (probability, localisations, and impacts). It appears that local levels of government (regions, provinces, and municipalities) have a heterogeneous understanding of their risks, lack analysis of potential disaster damages and have limited risk maps. As a consequence, the national risk atlas is not sufficiently consistent across the country, and it does not include damage analysis.

...and access to risk-related information should be the subject of an explicit data policy

The observed lack of coordination and data-sharing among data producers goes hand-in-hand with data-access difficulties for users. A large part of the risk information is either of scientific nature -with a research perspective - or of operational nature, in a dedicated tool owned by a public agency. This causes all or part of the risk information databases to be inaccessible to public decision-makers (such as urban planning agencies) or to private decision-makers (citizens and investors).

Therefore, Moroccan risk assessment needs to be formalised

The risk assessment process seems incomplete because numerous key players lack accurate risk information to make informed decisions, despite the development of efficient but under-used tools, especially at local level. It appears, therefore, that efforts made have not been replicated elsewhere, that the tools put in place have not been regularly updated, and that the appropriation of developed risk assessment tools has remained low. This points to a certain number of deficiencies in the risk assessment process in Morocco that need to be resolved; for example, the Ministry for the Environment's risk atlas does not cover vulnerabilities and potential damage.

In particular, deficiencies in risk assessment and risk mapping manifest themselves at local level, as they fail to inform policies relating to town/local planning, the development of risk scenarios for planning emergency resources, and the development of corresponding emergency response plans. Generally speaking, a risk assessment process should make all existing measures consistent and ensure that all risk information needs are met.

Three approaches developed in OECD countries could inspire Morocco

- The creation of a loose partnership between the various scientific and technical institutions, based on the Natural Hazards Partnership model in the United Kingdom.
- The development of an independent risk observatory – like the National Risk Observatory in France – that would bring together data providers and information and users, including private companies, regional authorities and scientific institutions.
- The establishment of a dedicated technical institution within the Ministry for the Environment, the Ministry of General Affairs and Governance or the Ministry of the Interior that would provide risk assessment expertise and information, based on the CENAPRED model in Mexico.

Regardless the selected institutional plan, the objective would be for the newly created body to be responsible for the risk assessment process in the country, collect risk-related data and information and make them available, develop methodologies and guidelines to ensure that risk assessments are carried out at different levels, and support the creation of an open and transparent national dialogue about risk, acceptable levels of risk, and the development of related standards.

Recommendations

ESTABLISH A NATIONAL RISK ASSESSMENT POLICY

- Establish a National Risk Observatory, or an equivalent body.
- Combine university and technical expertise, by mobilising the capacities of the entire Moroccan scientific community for assessing the risks through a public research programme.
- Develop national risk assessment standards.
- Develop risk mapping at the local level, with up-to-date records and capacity building programmes in tandem with local authorities.
- Draft an explicit data-sharing policy after national expertise and tools have been strengthened.

Risk prevention: Additional efforts

Risk prevention: A developing policy in Morocco

For many years, investments in structural prevention measures have been primarily concentrated on water infrastructure. An examination of Morocco's risk prevention expenditure across various programmes shows that the vast majority tends to finance water-related infrastructure (approximately 90% of expenditure). A large dam policy, adopted in the 1960s, was aimed at developing irrigation, as well as other water uses. Since then, several large dams have also been built following significant flooding.

This policy continues to this day; the construction of seven new dams has recently been launched, while the National Water Plan's investment plan runs until 2030. The policy is also consistent with the significant investment planned for the 450 km north-south water transfer, which would feed the river basins under water stress for the Rabat, Casablanca and Marrakesh regions.

Flooding was an issue of serious concern in the 2002 National Flood Prevention Plan, which had identified 391 at-risk areas, including 50 priority locations. This programme had a significant investment plan for structural measures for both containment and embankment protection.

Today, questions are emerging about the limits and the marginal returns of the dam building policy. Building dams tends to be increasingly costly as are maintenance costs. The potential effects of climate change brings deep uncertainties regarding water resource availability in the future, and the efficiency of dam building for flood management can also at times be questioned.

Non-structural measures are diverse but limited in scope

Morocco's risk culture has long been based on the memory of past events such as the Agadir earthquake in 1960, which is enshrined in the collective memory of populations living in the area. However, this collective memory remains weak in sites that have not suffered recurrent disasters and in rapidly urbanising cities. Awareness, information and communication about risk have been neglected, despite a number of pilot programmes; opening the civil protection barracks on March, 1st every year, distributing brochures in schools, running a seasonal forest fire campaign, raising awareness among elected officials. There is still no regulatory framework that requires local governments to inform local populations about their potential risk exposure.

A critical issue in risk prevention relates to urban planning and land development: existing urban planning documents do not include the question of risk beyond basic information. The Urban Development Master Plans are long-term references for urban planning at the local level, which were developed in the absence of risk mapping in many cases. Although the Hydraulic Basin Agencies and the Urban Agencies provide mandatory notices to local government, the lack of precise and legally valid risk mapping remains a key obstacle to the implementation of non-structural preventive measures.

Making local stakeholders responsible is necessary to improve prevention policies…

Strengthening prevention through co-construction and responsibility sharing across stakeholders at the local level will help move beyond structural measures, which are costly and are now reaching their limits. Non-structural measures shared locally by interested parties are an essential condition to implement prevention efforts, while moving beyond current top-down approaches. If such a locally-owned approach starts to emerge through policy instruments (such as the community action plan or the FLCN call for proposal), it will be important to formalise it further, as well as to increase incenticves and improve control for the effective implementation of local prevention policies.

The current process of regional reform and decentralisation offers new opportunities for an improved integration of risk management into urban planning documents.

…as is increased awareness and clearly articulated responsibilities for individuals and businesses

Strengthening the risk culture by raising awareness, running campaigns targeting populations, businesses and local agencies, is a key way to foster a locally-owned risk prevention approach. These programmes could be developed within a national campaign to train public officials and to ensure that risk policies are well understood nationwide. Further training opportunities within the education system, the media and existing associations should also be part of an ambitious awareness policy.

Business resilience does not only involve large systems and infrastructures. The location of industrial districts within high-risk areas in Casablanca resulted in losses for almost 200 industrial companies of all sizes in 2009. Since relocation is a very costly proposition, a prevention policy would be more effective, most importantly for small and medium enterprises. Initiatives led by professional organisations, such as conferences on risk management organised by the General Confederation of Businesses in Morocco, should be strengthened. However, economic development initiatives led by public authorities (e.g. the promotion of industrial districts in the Industrial Development Plan) have largely ignored the question of risk.

The significant impact of flooding on critical infrastructure

In November 2014, almost all the cities of Sidi Ifni province became disconnected from the national road network following floods. This complicated significantly rescue efforts and forced authorities to organise supply by sea. The risk of drought can also have consequences on the water supply. This was the case during the 1995 drought, when the city of Tangiers had to receive supplies by tanker ships. Despite considerable efforts by the Ministry of Public Works, Transportation and Logistics, an increase in the levels of prevention and infrastructure network coordination is still needed to strengthen resilience and prepare for critical risks.

Beyond critical infrastructure, a specific policy should be dedicated to the resilience for registered industrial installations, where accidents could lead to large disasters. The regulatory framework for these installations is determined by a practically unchanged 1914 *Dahir* (King's decree) regulating unhealthy, impractical or dangerous establishments (Official Bulletin n°97 of 7 September 1914). This is causing Morocco's risk studies, certification procedures, and inspections and monitoring of installations to be inadequate. Examples of policies supporting critical infrastructure resilience in Finland, France, the United States and the United Kingdom could serve as models for Morocco on this subject.

The risk of tsunami is still insufficiently accounted for

A tsunami could have significant consequences for Morocco and cause long-term negative effects on its economy and society. Tsunamis are rare, but a previous tsunami had major impacts on Portugal, Spain, and Morocco (1755). Today, given the importance of coastal activity, a tsunami would be the worst-case scenario for Morocco. Smaller-scale events, such as strong sea swells, should also not be overlooked, as their consequences could be significant for ports and marinas. A discussion on this risk raises the question of the ocean's role in Morocco's future developments. At what point and under which conditions should the country's economic development be based on the ocean economy? And what is an acceptable level of risk?

In the medium and long term, prevention financing remains vulnerable until it is made permanent

The total amount of prevention expenditure in Morocco is difficult to estimate because financing sources are diverse: sector-based plans, infrastructure, and funds specific to ministries all provide prevention funding. A short-term extension of resources will be made possible by a loan from the World Bank for 2016-2021. At the same time, strengthening non-structural prevention measures also depends on decisions that are weighed against structural measures. In the longer term, balancing resources with prevention needs could prove difficult, and more permanent financing should be sought, specifically for the Fund to Combat the Effects of Natural Disasters (FLCN).

Recommendations

ESTABLISH A PERMANENT PREVENTION POLICY TO IMPROVE THE CURRENT STATE OF PLAY AND STRENGTHEN LOCAL PROGRAMMES

- Preserve and reinforce prevention financing via the FLCN, prioritise funding and activities and emphasise the dissemination of best practices.
- Improve the process by which priorities are balanced between structural and non-structural activities to optimise investment and the use of public funding.
- Initiate a national risk communication campaign focused on local responsibility.
- Strengthen private sector preparedness.
- Improve integration of resilience into regional and urban development through transparent and binding zoning policies.
- Establish a partnership with critical infrastructure operators, and include them in risk management.
- Reflect more actively on prevention related to the risk of tsunami.

Emergency management: Shift into overdrive

A system that is developed progressively and adapted to local emergency management needs

Increased attention has been given to emergency management since 2000, as disasters demonstrated the inadequacy of funding, limited capabilities and delayed emergency intervention. Civil Protection has seen its funding and level of professionalism strengthened. The creation of the Monitoring and Coordination Centre (CVC) in 2008 under the Ministry of the Interior strengthened the existing framework by providing a mechanism for crisis management and decision-making in real time.

This system is based on subsidiarity and a centrally coordinated chain of command with regional representatives. The Civil Defence Administration prepares the Master Plan for Risk Analysis and Coverage (SDACR), which plans emergency resources and capabilities for the national and regional levels, based on its risk analysis. Emergency response is based on a generic multi-actor Emergency Response Plan that also includes regional specificities with specific plans for specific risks. Warning, communication and information sharing systems are gradually improving, thanks to better coordination. The capacity to scale-up response, inter-agency coordination and cooperation, is being strengthened with high-level national centres (e.g. INSARAG certified) to support vulnerable sites on a more regional basis.

In the meantime, emergency response capabilities remain too limited to face large or very large-scale crises

Civil Defence human resources (approximately 8,000 people, or 0.25 per 1,000 individuals) are still relatively limited and it is often necessary to call on the Royal Armed Forces and other security forces for assistance in critical risk situations. The Ministry of Public Works plans for and uses its own means of intervention to ensure the continuity of infrastructure services for which it is responsible. The health sector, despite contingency plans, seems largely under-prepared if a critical risk were to occur. In this context, coordination between organisations is essential to ensure areas that are under-prepared do not weaken the response overall effectiveness.

Some critical risks remain insufficiently grasped

Just as a longer time horizon must be sought for risk assessment, emergency planning efforts should consider scenarios of lower probabilities and higher potential damages. It seems there is not yet an emergency plan for the risk of a tsunami, for a significant earthquake affecting a large portion of northern Morocco or for two simultaneous large floods in two widely separated areas of the country. These more extreme risks regularly occur in other countries, and Morocco cannot afford not to plan for them.

Warning systems have been strengthened but coverage remains insufficient…

For example, for floods, only 60% of catchment basins are equipped with a telemetry system to measure water depths and flow rates for meteorological flood warnings. The system of tidal gauges and buoys for surges and tsunamis is not sufficient, nor is the integration of real-time satellite images for monitoring and warnings.

A unified information and telecommunication system for emergency response agencies with a dedicated channel and the use of social media to inform individuals are additional ways to improve crisis communication. Social media is widely used in Morocco, which creates a strong potential for further improvements in this area.

…justifying increased involvement of civil society and communities

Crisis simulation exercises are driven by strategic, tactical or operational plans, particularly with international partners. However, they are infrequent and not sufficiently inclusive. Morocco has a high volunteer capacity for emergency services due to a strong associative culture. Mobilising this volunteer spirit and adopting a positive attitude to emergency services would help improve the country's risk management culture. This capacity should be cultivated and strengthened, including through the clarification of volunteers' status, training and certification. International partners have already shown their readiness to support these efforts.

On a qualitative level, vulnerable groups are likely to become more diverse in the coming years. These groups, including the elderly, chronically ill, handicapped, tourists, and in-transit immigrants, will require special assistance and specific responses.

Recommendations

RAISE THE LEVEL OF EMERGENCY PREPAREDNESS FOR MAJOR CRISES

- Strengthen the Monitoring and Coordination Centre (CVC) by coordinating with warning systems and critical infrastructure.
- Pursue the coordination of capabilities and strengthen emergency response capacity in accordance with the results from the national risk assessment.
- Continue to strengthen the country's early-warning system, specifically for floods and tsunamis.
- Develop inter-agency emergency plans based on principal risk scenarios at the national and regional levels.
- Develop an active strategy for crisis information and communication using new technologies, particularly social media.
- Encourage the development of volunteer groups for crisis management by developing the status of volunteers.
- Strengthen international cooperation activities in emergency response.

Recovery and reconstruction: Should financing be consolidated?

How can the traumatising consequences of major events be limited?

Minimising the impacts on individuals and businesses, in economic, social and psychological terms, largely depends on how quickly services can be restored. Medium and long-term traumatic consequences for individuals, families and businesses will be minimised based on the speed and adaptation of the response. However, speed should not be synonymous with improvisation. Important decisions must be made in real time and must therefore be adequately anticipated. The risk of inadequate planning is that those affected (individuals, businesses, municipalities) rebuild in the same way and in the same location in the midst of a flood or seismic zone. Since 1960, Morocco has demonstrated in Agadir that it is taking this particular issue into account. It is also one of the major challenges that prompted the promotion of the "Build Back Better" concept by the OECD and the United Nations. In addition, this is one of the potentially "positive" aspects of destruction/reconstruction, as it can lead to better, more sustainable and less vulnerable buildings.

Governments are often at the forefront of the reconstruction process, including rebuilding public infrastructures. Gradually, Morocco has been able to move from a reactive response (budget amendments, requests for international aid) to the creation of dedicated mechanisms and insurance solutions with or without the market.

...By taking the first steps toward the implementation of financing mechanisms...

The establishment of the Fund to Combat the Effects of Natural Disasters (FLCN) in 2009 allowed the government to provide funding for emergency response, recovery and reconstruction. The government contributes 200 million Dirhams annually to the fund, which is also used for prevention efforts. The flexible rules regulating the use of this fund provide financing based on the type of emergency, reconstruction and prevention needs.

A more streamlined approach to available financial instruments for reconstruction or compensation has also been established, along with the development of insurance schemes. Mostly consisting of government subsidies for the agricultural sector, these insurance mechanisms provided Morocco with experience in building tools that combine both principles of mutual insurance for those who can access them, and of aid for those who are unable to access insurance.

...By implementing innovative approaches with the assistance from insurance mechanisms

This combination of insurance and aid principles can now be extended (Law 110-14) to individuals and their property. This Law also extends the compulsory insurance spectrum. As it is the case in other countries, the long-term benefits of such a policy will occur when they encourage behavioural changes, while avoiding the moral hazards that can be caused by providing too much assistance.

...And taking into account the continuity in business activity...

Similar to other countries, Morocco's informal sector, including craftspeople, merchants, transportation companies, small industries or subcontractors, and small service companies are often poorly prepared and insured for natural or industrial disasters. This issue has already been examined by the Central Guarantee Fund and a section of the Fund to Combat the Effects of Natural Disasters.

Concerning the financial contingencies borne by the government and its various ministries, particularly pertaining public infrastructure damage, an integrated risk management strategy could also include dedicated financing tools and the possible transfer of such risks. Ultimately, this would settle the question of establishing a system dedicated to the risk of significant losses for the government's budgetary resources in the event of a major disaster.

Better define the financial responsibilities of public financing in advance.

What amount of compensation should be included in a budget amendment compared to a fund to be annualised to allow absorption of prevention and reconstruction costs?

Part of the answer lies in the amount of spending the Moroccan government channelled to finance recent disasters, and the jolts these costs have triggered. The use of the MnhPRA model for assessing damage to public infrastructure could allow different hypotheses to be tested to evaluate the suitability of various mechanisms for covering damage.

Recommendations

CONSOLIDATE AND STREAMLINE NATURAL DISASTER FINANCING SYSTEMS TO PROMOTE RESILIENCE

- Establish permanent public financing for critical risks.
- Use evaluation to better estimate future financial needs.
- Pool budgetary mechanisms for compensation, using clear rules established in advance concerning the use of public funds.
- Consolidate and extend the agricultural sector's insurance system, and accelerate the implementation of coverage mechanisms for individuals.
- Promote "Build Back Better".

Chapter 1.

Principal risks in Morocco

Morocco is exposed to a variety of risks of natural, industrial or human-induced origin. Among these risks, the greatest in terms of potential impact on the human population and economy, are floods, drought, earthquakes and tsunami. These risks could affect numerous regions across the country. This chapter provides an overview of the main risks threatening Morocco and it examines their corresponding socio-economic vulnerabilities. Factors aggravating these vulnerabilities – such as climate change – must be taken into consideration in the future. These risks are therefore of growing concern for Morocco.

Introduction

The diversity of the Moroccan landscape, the variability of climate and weather conditions, and the country's geography and geology all contribute to exposing its territory to a number of natural hazards. These can take the shape of floods, drought, earthquakes, tsunamis, forest fires, heat waves and cold snaps, as well as storms and submergence, landslides, locust infestations and blizzards. A significant proportion of Moroccan inhabitants and several key sectors of the country's economy are particularly vulnerable to these risks. The most critical ones in terms of potential impact on the human population and the economy are floods, droughts, earthquakes and tsunamis, which could affect numerous regions across the country. This study, therefore, focuses on these four specific risks of disaster.

Major shocks in OECD countries and beyond have shown the significance and impact of natural disasters. The 2011 earthquake and subsequent tsunami in Japan, Hurricanes Katrina and Sandy in the United States, the earthquakes in Chile in 2010 and in New Zealand in 2012, and the severe floods that affected Central Europe in 2011 and Queensland (Australia) in 2011 were all major catastrophes whose effects were not sufficiently anticipated. In Morocco, the tsunami of 1755, a consequence of the Great Lisbon Earthquake, the 1960 Agadir earthquake and the 2004 Al Hoceima earthquake constituted major disaster events. The human cost of these disasters, the losses and damage suffered by households and businesses, their negative impact on government budgets, the tax-related and macroeconomic repercussions, sometimes extending beyond national borders, and their impact on the environment and on the cultural heritage are all key challenges that governments should address by crafting and implementing appropriate public policies.

In their absence, or in cases where existing public policies and mechanisms do not meet these challenges, the effects of catastrophes end up costing even more and can lead to the erosion of trust in governments and institutions, particularly given citizens' considerable expectations in that sphere. Public policy reforms often follow disasters for which governments were not sufficiently prepared in advance, despite the fact that better foresight could have enabled them to cope with them through targeted policies.

Against a socio-economic backdrop marked by high uncertainty, ensuring Morocco's resilience to critical risks is a fundamental concern. While the country's development model and the political reforms undertaken during the past decade have enabled Morocco to enjoy a noteworthy acceleration in economic growth and an improvement in population well-being, the country is still facing numerous socio-economic challenges. A major shock could adversely affect the aforementioned trajectory and have negative repercussions for the population's well-being, the country's stability, as well as for public trust in its institutions. Morocco has experienced regular catastrophes of significant socio-economic impact. Floods, drought, and earthquakes (although these are rare) have all had consequences for authorities. For example, the 2004 Al Hoceima earthquake caused 628 deaths and 926 injuries, while causing an estimated USD 400,000 of damage. The drought of 1994-1995 hampered the crop year, reducing annual grain production by over 70%. Agricultural GDP subsequently dropped by 40%, leading to a 12.4% overall reduction in GDP in 1995. Finally, the floods that struck Guelmim and Sidi Ifni in November 2014 resulted in 47 deaths and 150 destroyed homes (EM-DAT).

This introductory chapter outlines the key issues relating to the main disaster risks in Morocco by examining in greater detail the ones that this study is focused on. It also attempts to present the main economic and societal vulnerabilities vis-à-vis these risks, and raises questions about the long-term trends that influence risk factors in Morocco.

Natural hazards in Morocco

According to the historical records of extreme events collected in the international EM-DAT database of the Centre for Research on the Epidemiology of Disasters (CRED) at Louvain Catholic University, 47 natural disasters struck Morocco between 1960 and 2015 (Table 1.1). While this is not an exhaustive database covering all disasters that occurred during this period[1], the table below provides a summary of various risks of natural origin and their impact.

Table 1.1. Principal natural disasters in Morocco, 1960-2015

Risk	Number of events	Deaths	Total population affected	Estimated costs (in USD thousands)
Drought	5	0	412,000	900,100
Earthquake	3	12,728	38,465	520,000
Flood	32	1,682	638,455	330,200
Storm	3	50	117,000	300,050
Extreme temperature	2	0	7,500	809
Pandemic	1	200	2,942	0
Landslide	1	1	12,216	0

N.B. The international EM-DAT database used here includes disasters that meet one of the following criteria: a) Deaths of at least 10 people, b) affected at least 100 people, or c) official declaration of a state of emergency or disaster, and d) launch of an appeal for international aid.

Source: EM-DAT, Louvain Catholic University, 2015

70% of these events are floods, which is the type of disaster that has affected the highest number of Moroccans. Periods of drought have caused more economic damage, while earthquakes have led to the highest number of deaths, particularly the Agadir earthquake of 29th February 1960, which had a magnitude of 5.7 and killed 12,000 people. During this period, Morocco was also affected by storms, extreme temperatures (heat waves and cold snaps) and landslides. Morocco must also cope with the risk of wildfires, particularly forest fires. These are not included in the database but occur, nevertheless, at an estimated rate of 285 per year on average, according to the High Commission for Waterways and Forests and Fight against Desertification (HCEFLCD, 2014). The risk of tsunamis should also be monitored closely. Although no examples of destructive tsunamis have been recorded during the period under consideration, the 1960 Agadir earthquake is said to have produced a tidal wave that penetrated 275 metres into the town.

The following sections outline the respective characteristics of each examined critical risk in Morocco in terms of potential socio-economic impacts, namely floods, drought, earthquakes and tsunamis.

Regular flooding affects many different regions in the country

The risk of flooding threatens numerous regions in Morocco. The river basins boast extremely diverse geological and hydrological characteristics, due to a highly heterogeneous climate and Morocco's general geography. Yet, they are all characterised by high rainfall variability that depends on the year. The water regimes of the country's rivers are notable for alternating wet and dry periods with high run-off years, severe droughts, and major thunderstorms (Ministry of Energy, Mining, Water and Environment, 2012). The inhospitable terrain and steep-sided valleys of the Rif Mountains, the Middle Atlas, and the High Atlas (peaking at 4,165 m) also create the conditions for particularly reactive hydrology during periods of abundant rainfall.

A distinction can also be made between, on the one hand, flash floods in small coastal watersheds and torrential floods in mountainous areas, (both of which are difficult to predict and potentially have destructive socio-economic impacts) and on the other hand, more frequent floods witnessed in plain basins, at the foot of mountain ranges. The latter tend to affect more urbanised areas, thereby causing more significant economic damage. The extremely high flow variability of Moroccan rivers, many of which having dry riverbeds for large parts of the year, does not help raising awareness levels among the general population. A number of buildings are located on these riverbeds and the rivers' natural floodplains, particularly in fast-developing urban areas.

Although floods struck the country and caused serious damage in 1950 in the Sefrou Region, in 1963 in Moulouya and in 1965 in the Ziz Valley, the last two decades have been marked by more frequent national flooding. In particular, more regular large-scale flooding occurred during the 2000s (Table 1.2). Increased human occupation of vulnerable areas and soil sealing linked to urbanisation, as well as the loss of natural plant cover caused by increased climate variability, are the underlying factors explaining this trend (Ministry of Energy, Mining, Water and Environment, 2008).

These floods can affect isolated rural areas in the south (47 deaths in November 2014 in Guelmim Region, several hundred deaths in Ourika valley close to Marrakesh in 1995 – see Box 1.1) and in the large urban centres in the north (in 2002, floods in Mohammedia, Berrechid, and the Greater Casablanca, killed 63 people, ruined hundreds of houses, and damaged the country's largest oil refinery; in 2008, 30 people died in Tangiers and the industrial district was badly hit). Regions can also be affected more broadly by river flooding, where the spatial extension and extended length result in major economic consequences. For example, flooding from the Sebou River significantly affected the Gharb Valley, Morocco's main area of agricultural production. It led to the deterioration of almost 100,000 hectares of farming land, as well as severe financial losses estimated at nearly 1bn Dirhams.

This review of flood events in Morocco underlines the diversity of this risk, which can affect just about every region in the country. It can have major impacts on human populations in cases of flash floods in small watersheds that are extremely reactive. The economic damage, which is rarely thoroughly assessed, can be quite substantial. It includes, in particular, the impact on infrastructure networks (roads, railways, bridges, electricity, sanitation and drinking water) and the damage borne by the agricultural sector and industrial districts. Furthermore, indirect damage linked to the disruption of infrastructure networks, for example, can also negatively affect economic activity and lead to additional economic losses. A model estimates the average annual losses incurred by floods in Morocco to be 4.2 bn Dirhams, with an impact that could rise as high as 27.5

bn for a once-a-century event and 34 bn for a once-a-millennium event (World Bank, 2014).

Box 1.1. The Ourika Valley flood of 1995

The Ourika flood of 17 August 1995 is part of the collective memory of disasters in Morocco because it resulted in major losses in human life; the official count was 150 deaths and 88 people unaccounted for. Ourika Valley features an *oued* with a very reactive flow, due to its steep slope and the rocky, impermeable soil of the majority of this very mountainous watershed. During the flood of 1995, the flow created a sheet of water several metres high that devastated the villages in the Valley, destroyed numerous houses and public infrastructure, and wreaked havoc with farmland. The rather high death toll that resulted from this catastrophe can be explained by the fact that it is a very popular summer destination for tourists, who are attracted by its fresh air and proximity to Marrakesh. The Valley features just one road running the length of the *oued* and very few places to take shelter in the event of such a disaster, which took people in the area – unaware of the risk – by surprise. In the aftermath of the flood, it took several years for the tourism industry to rebound in the Ourika Valley, where 5,000 cars arrive daily at the peak of the season.

Source: Tensift Hydraulic Basin Agency (2010), Flood risk atlas; EM-DAT database, Louvain Catholic University, Belgium, interviews carried out as part of the OECD Study on Risk Management in Morocco (May 2015).

Table 1.2. Main floods in Morocco (1995-2015)

Year	Region affected	Town/area affected	Deaths	People affected	Damage (in USD thousands)	Other damage
1995	Marrakesh-Safi	Ourika Valley	730[2]	35,000	9,000	142 buildings destroyed, 300 ha of farming land flooded, road damaged, bridges destroyed
1995	Fez-Meknes	Taza; Oued Amlil; Zouagha	43	-	-	
1995	Souss-Massa	Tata	18	3,000	-	
1996	Beni Mellal-Khenifra	Beni-Mellal	25	60,000	55,000	
1997	Fez-Meknes	El Hajeb	60	-	-	
2000	Tangier-Tétouan-Al Hoceima	Martil	6	650	-	
2000	Fez-Meknes	Taza	0	300	-	
2001	Casablanca-Settat; Marrakesh-Safi	Settat; Essaouira	15	300	2,200	
2002	Casablanca-Settat	Mohammedia; Berrechid	63	15,000	200,000	17 industrial units damaged (to the tune of 30 million Dirhams), fire at the Samir refinery, where work came to a standstill for several months (losses totalled 1.5bn Dirhams), roads cut off, farming land flooded, loss of livestock, houses collapsed, hundreds of homes flooded, hundreds of hectares of farming land affected

Table 1.2. Main floods in Morocco (1995-2015) *(continued)*

Year	Region affected	Town/area affected	Deaths	People affected	Damage (in USD thousands)	Other damage
2003	Tangier-Tétouan-Al Hoceima; Oriental	Bni Boufrah; Nador; Al Hoceima	35	10,000	-	
2006	Drâa-Tafilalet	Ouarzazate; Errachidia	N/A	-	-	
2006	Marrakesh-Safi	Essaouira; Safi	11	2,100	-	
2006	Drâa-Tafilalet	Errachidia	6	-	-	
2008	Tangier-Tétouan-Al Hoceima	Tangier	30	20,000	-	Industrial district devastated
2008	Marrakesh-Safi	Marrakesh	9	0	-	
2009	Rabat-Salé-Kenitra	Gharb area	29	9,500	-	400 houses destroyed, 90,000ha crops ruined, infrastructure (water, electricity) affected
2010	Marrakesh-Safi	Essaouira	32	75,000	29,000	
2014	Guelmim-Oued Noun	Guelmim	60	117,000	300,000	Roads, bridges and embankments destroyed, 150 houses also destroyed

Source: World Bank (2014), EM Data (2015), Ministry of Energy, Mining, Water and Environment (2015) and Ministry of Public Works, Transportation and Logistics (2015).

Risk of drought

Drought also constitutes a critical risk in Morocco, due to the important role the agricultural sector plays in the Moroccan economy. The agricultural sector contributed to 13% of the country's GDP in 2014 (according to the World Development Indicators). It employs 43% of the Morocco's labour force and 78% of the working population in rural areas (Ministry of Agriculture and Maritime Fishing, 2012). The agricultural sector is extremely dependent on rain, as only 15% of the farmland is irrigated (Figure 1.1). But total average rainfall, estimated at140bn m^3, varies from year to year, with a ratio of 1 to 5, from 50 to 250bn m^3. It only falls during a limited number of days, estimated at nearly 20 days in the south and 70 in the north (Balaghi et al, 2007). In these conditions, agricultural production is highly susceptible to the risk of drought caused by a shortfall in rainfall or poorly distributed rainfall (both in terms of time and space).

Consequently, Morocco suffers from regular spells of drought, typified by a deviation from the standardised precipitation index. During the twentieth century, Morocco endured 11 principal droughts, which affected the entire country and lasted for significant periods of time – up to six years between 1930 and 1935 and between 1980 and 1985 (Mokssit, 2008). Although the lack of regularity of these spells was noted, a number of authors also pointed to the increase in the number of drought years at regular intervals over the course of the century. These have risen from one drought year out of every five years to a drought year every second year, due to the effects of climate change (DMN, 1997; IRES, 2013).

The repercussions for the nation's economy can be extremely significant (Table 1.3): During the drought year of 1995, poor agricultural production led to a drop in national

GDP of 6.6 points, while the following year witnessed three times more rain and production undergo an increase of 78%, leading in turn to a rise of 12.2% in GDP (Balaghi et al, 2007). A World Bank assessment estimated the risk of drought would result in an average annual economic loss of 4.6 bn Dirhams in Morocco's grain production, corresponding to nearly 15% of agricultural GDP (World Bank, 2014).

Table 1.3. Examples of socio-economic impact of recent periods of drought in Morocco

Period of drought	Principal socio-economic impact
1981-84	Reduction in agricultural production of approximately 40%
1994-95	Fall in grain production of between 9.5 and 1.6 million tonnes Drop in GDP of 6.6% Major wave of rural migration and development of informal housing
1996-97	Drop in GDP of 2.3%
1999-00	275,000 people affected Economic damage estimated at USD 900 million. Wheat imports double (from 2.4 million tonnes to 5 million tonnes on average).
2004-05	Economic growth rate reduces by 3.5 to 1.3% for 2005
2006-07	700,000 people affected Halving of grain production

Source: Royal Institute of Strategic Studies (2013), *Review of extreme climate risks and their impact on the Moroccan economy*

Furthermore, drought incidents can also have an impact on the availability of water resources. With average resources of 730m^3 per inhabitant per year, Morocco suffers from a chronic water shortage. A country is considered to be experiencing water stress when resources fall under 1000 m^3 per inhabitant per year. This can have significant repercussions on non-agricultural water usage, especially for supplying drinking water to towns and for hydroelectric production.

As a consequence, in Tangiers, the country's third largest town in terms of population size, in 1995, Morocco was forced to charter four tankers to transport 5.6 million m^3 of water to supply its water systems and cope with water shortages in the dams supplying towns for seven months (IRES, 2013). The cost of this operation was estimated at 250 million Dirhams. Launched in the 1960s, the large damns policy contributed to the development of irrigation and to the mobilisation of water resources for other uses. However, tension surrounding water availability remains very high, as water scarcity and variability are exacerbated by a changing climate.

In addition to having a strong economic impact, periods of drought can also have major social effects by creating a considerable surge in urban migration. Rural populations, whose only source of income comes from farming work, are particularly vulnerable in times of severe drought, especially if it lasts for several consecutive years. Although the number of people living in Morocco's shanty towns decreased in the first half of the 1990s, the major droughts of 1994 and 1995, and the ones of 1999 and 2000, caused a natural exodus from rural areas. This saturated the capacity of towns to absorb the incoming urban stream, causing the sprawl of existing shanty towns and the appearance of new ones (AFD, 2011).

Figure 1.1. Variations in GDP, agricultural GDP and annual rainfall during agricultural campaigns, 1992 to 2011

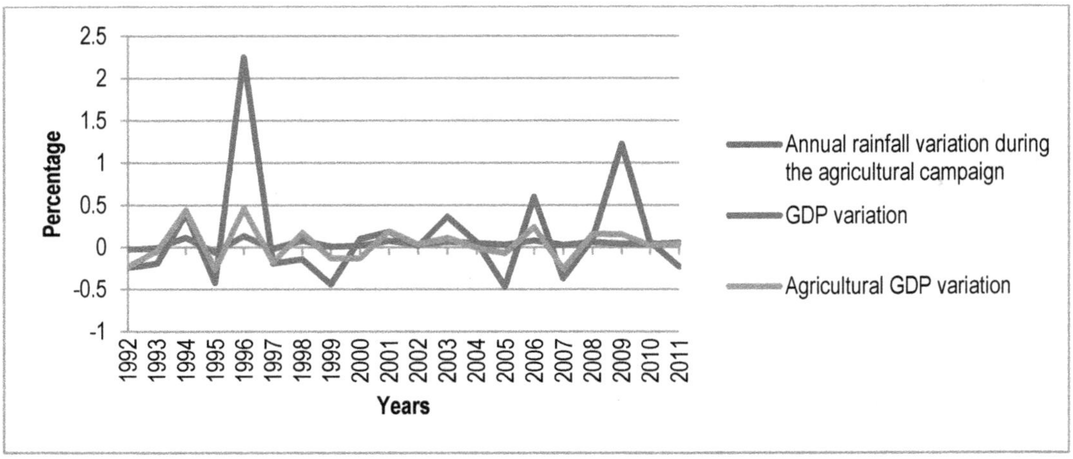

Source: Bank Al Maghrib (2011), World Development Indicators (2016)

Risk of earthquakes

The risk of earthquakes is more localised, particularly in the north of the country and in the Agadir Region. The 1960 Agadir earthquake (12,000 deaths and 70% of the city destroyed) and the Al Hoceima earthquakes of 1994 and 2004 (600 deaths and 12,000 homes destroyed) illustrate the high stakes of managing such potentially devastating disasters (Table 1.4). Despite their relatively moderate magnitude (5.7 and 6.3 respectively on the Richter scale), these earthquakes caused substantial damage, mainly due to the buildings' vulnerability. In the past, the 1755 Great Lisbon Earthquake (8.7 on the Richter scale) was the most damaging, as it hit a large part of Morocco's Atlantic coast, which was particularly affected by the tsunami that followed. The inner part of the country was also seriously affected by the aftershocks that struck in the cities of Fez and Meknes.

Table 1.4. Earthquakes in Morocco's recent history

Year	Town/area affected	Magnitude	Human casualties	Economic damage (in USD thousands)	Unquantifiable damage
1960	Agadir	5.7	12,000 deaths, 25,000 injuries	-	70% of buildings destroyed
1994	Al Hoceima	5.6	6 deaths	-	-
2004	Al Hoceima	6.3	628 deaths, 926 injuries	400,000	12,376 of houses collapsed in rural areas, 967 in urban areas 15,320 people left homeless

Source: World Bank (2014), Cherkaoui (2004) and EM Data (2015)

At crossroads between the Eurasian and African tectonic plates, Morocco is considered to have a moderate seismicity level in comparison to other Mediterranean countries in the region. A compilation of 27,500 earthquakes observed between 1901 and 2010 (Cherkaoui, 2012) demonstrates that the regions surrounding Al Hoceima in the north and Agadir are the most exposed areas to the risk of earthquakes, where these could reach destructive levels of intensity. The regions situated on the Tangiers – Essaouira coast, the inner part of the country around Fez, Meknes and Tafilal, and the Rif and Middle Atlas mountain range, display more moderate seismicity levels.

This generally moderate seismicity can nevertheless cause major damage due to Morocco's significant vulnerability to this risk, as has been noted during past events. This includes a low adherence to earthquake resistance standards, the strong presence of makeshift housing (e.g. at the foot of mountain ranges near the fault in the northern part of Agadir), and the specific features of certain towns that have densely populated historic centres (e.g. medinas), which complicates access in the event of a disaster. The World Bank estimates at 850 million Dirhams the annual average losses caused by the risk of earthquakes. These estimations expect once-a-century events and once-a-millennium events to cause losses of 15bn and 34bn Dirhams respectively (World Bank, 2014).

Risk of tsunamis

The Moroccan coast is also vulnerable to the risk of tsunamis, such as the one that hit in 1755 following the aforementioned Great Lisbon Earthquake. Historical sources list a dozen earthquakes that could have created a tsunami capable of reaching the Mediterranean and Atlantic coasts of Morocco. Among these, Morocco's tsunami records shows six that definitely resulted in tsunamis in 382, 1755, 1761, 1941, 1969 and 1975 (Kaabouben et al, 2009). The 1755 tsunami remains the most damaging for a large section of the Moroccan coastline (Box 1.2). Other lower-intensity tsunamis have also occasionally hit Morocco, such as the one in 1969. It was generated by an earthquake with an epicentre located in the same zone as the 1755 earthquake, but which was much less severe (7.3 versus 8.7 on the Richter scale). Consequently, the waves did not surpass 0.9 m and the damage incurred was relatively minor (IAU, 2011). While the likelihood of a destructive tsunami occurring in Morocco remains low, it is important for the country to take this type of risk into consideration, given the potentially high damage it could cause today. Probability estimates carried out by the World Bank put economic losses of a once-a-millennium event at 58bn Dirhams (World Bank, 2014).

Box 1.2. Tsunami of 1755

An interpretation of historical archives and sediment analyses provides an idea of the damage that the 1755 tsunami caused in Morocco, although estimates are sometimes inconsistent and can include some of the damage instigated by the earthquake itself.

Thus, major damage was recorded along Morocco's Atlantic coastline, particularly in the towns of Tangiers, Asilah, Salé, El Jadida and Safi. Depending on the source, the wave height is said to have varied from 2.5m to 4m in the majority of those towns, with a few mentions of higher 15m waves in Safi and Tangiers, although this estimate is considered as exaggerated by some authors. The damage is sometimes described as having largely spread into the inner part of the country, up to 2km in the cases of Tangiers and Salé.

Simulations of this type of tsunami forecast that waves measuring 2m to 10m could reach the Moroccan coast within 30 minutes to an hour following an earthquake. The return period for

Box 1.2. Tsunami of 1755 *(continued)*

these various tsunami scenarios is estimated at 1,000 to several thousand years.

Source: Samira Mellas, *Assessment of the risk of tsunamis on Morocco's Atlantic coastline.* History. Paul Valéry University - Montpellier III, 2012. French. <NNT: 2012MON30067>, doctoral thesis.

Other major natural risks

Other risks of natural origin can have considerable impact on Morocco. The Mediterranean climate is conducive to the risk of forest fires, which regularly affect the country's stretches of forest, during the summer season. On average, 451 fire outbreaks are recorded every year (2006-2015). According to the High Commissioner for Water and Forests and Fight against Desertification, these are mainly concentrated in the northern part of the country. Approximately 3,000 ha are affected by forest fires annually, with repercussions for soil stability, erosion and consequently on the risks of floods and drought.

The northern part of the country is also exposed to the risk of landslides because of the soil types and their steep and hilly slopes, and the significant ground instability recorded there. In 2008, such an event in the Fez region caused the death of 52 people and destroyed several homes. The road network is also regularly affected by the impact of this type of risk (IRES, 2013).

Cold snaps and snowfalls regularly occur in the mountainous regions of the country, where temperatures can drop to -10 degrees Celsius for several days. Isolated villages and houses are not accessible to emergency services in the event of major snow fall. The exceptional cold snap of January 2007 caused nearly 30 casualties, including young children in isolated villages located in the Atlas Mountains, such as Angfou in the High Atlas. In January 2009, nine people died and villages were cut off from the road networks for several days.

Violent storms and sea swells also represent a risk that can have major economic repercussions on Morocco, particularly on maritime activities and business, while causing damage to coastal housing. The last extraordinary sea swell that occurred in January 2014 featured waves of a maximum height of 13 metres. It resulted in substantial damage to the port infrastructure (Ministry of Public Works, Transportation and Logistics, 2014). In particular, seawalls were destroyed in Mohammedia, the country's main industrial port. In the future, these risks are only likely to increase due to the long-term sea-level rise caused by global warming.

The risk of locust outbreaks is also a non-negligible risk for Morocco's agriculture. If the weather conditions are favourable, swarms of desert locusts present in the Sahelian Region are able to multiply very rapidly and migrate across all of West and North Africa in search of food. The outbreaks of 1987-1989 led to major losses, while the control treatment cost 1bn Dirhams. In 2003-2004, the treatment of about 3 million hectares was needed to prevent another locust swarm and the ensuing crop destruction.

Human-induced risks

Finally, risks of human origin should also be taken into account in an integrated all-hazard and threat risk management policy. These include industrial risks. In November

2002, the Samir refinery fire demonstrated the link that exists between natural risk and industrial risk. Floodwaters originating from Oued Mellah Creek in Mohammedia lifted waste oil and brought it into contact with hot parts of the refinery. This caused the start of a large fire that killed two people and injured four others.

In addition, Morocco also faces the risk of epidemics. In 2015, the Ebola epidemic forced Morocco to give up hosting the African Nations Cup football tournament. The risk of terrorist attacks, which has increased in light of the current difficult geopolitical context, could also have a damaging impact on Morocco, especially on its tourism industry, as was demonstrated by recent attacks striking countries in the region.

Effects of climate change

Climate change contributes Morocco's increasing vulnerability to disaster risks in several ways. Models from the National Meteorological Office (Mokssit, 2012) project simultaneously an increase of average summer temperatures by 2 to 6 degrees and a reduction of average rainfall by up to 20% by the end of the century. Furthermore, a vulnerability study on the effects of climate change estimates that by 2020, Morocco would experience an average reduction in rainfall of 11%, an imbalance in seasonal precipitation, and a decrease in snow cover duration and volume (Hulme et al, 2000). The risk of drought and an increase in the risk of forest fires are also expected. However, observations of recent changes in the country also show a perceptible rise in autumn rainfall that could lead to an increase in the risk of flooding because autumn is also the time of year with the highest levels of rainfall (Barkhordarian et al. 2013). Furthermore, the rise in sea level may influence the risk of storms and sea swells, resulting in an increase in wave height. The risk of tsunamis could also increase; the International Panel on Climate Change has predicted that sea levels will rise by an average of 28 to 98 cm by 2100, depending on different greenhouse gas emission scenarios (IPCC, 2013).

An IRES study (Messouli, 2013) illustrated the phenomenon of changes to the probability of hazard occurrence in Morocco for the coming decades by using examples of cold snaps and heat waves and making various climate change assumptions (Figure 1.2). In the event of an increase in frequency and severity of heat waves, a series of adverse effects are therefore to be expected in terms of agricultural crops, health of fragile communities, network disruption, electric consumption peaks, etc.

Figure 1.2. Frequency and length of heat waves

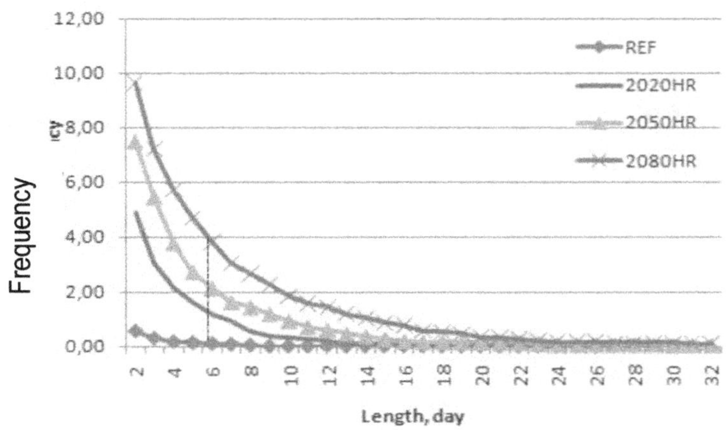

Source: Messouli (2010)

Socio-economic vulnerabilities

A variety of vulnerability factors specific to Morocco can heighten the potential impact of these different hazards. The concentration of strategic sites, key economic sectors and inhabited centres in high-risk areas, as well as the population's social vulnerability, need to be taken into account when analysing the country's risk factors.

Concentration of people and production in high-risk areas

While the risk of drought affects all the country's rural areas, the risks of flooding, earthquakes and tsunamis tend to be more localised. The demographic and urbanisation dynamic prevalent since the 1960s has made Morocco a predominantly urban country, where communities settle in coastal areas. According to the latest census in 2014, 60.3% of Morocco's 33.8 million people live in an urban environment. In ten years, the total increase in the population has been absorbed by cities, which grew by 4 million between 2004 and 2014. This population is largely concentrated in the coastal belt, especially in the north-south strip that runs from Tangiers to Agadir (representing 46% of the urban population), and in the country's great valleys (High Commission for Planning, 2014).

Morocco's population projections for 2030, produced by the High Commission for Planning, indicate a population increase of 10 to 20% by 2030, i.e. from 34 to between 38 and 42 million inhabitants, based on different fertility rate assumptions (Table 1.5). This would represent an average increase of 300,000 inhabitants per year and would entail a transition from 16.4 to 24.4 million inhabitants in the cities. Supposing an equal occurrence of hazard probability, this implies a significant worsening of risk exposure and harmful consequences.

Major urban population increase and sound urban development policies have not always gone hand-in-hand. New settlements are often concentrated in high-risk areas, such as those near *oued* beds, those in lower-lying areas, or at the foot of mountain ranges or in steep areas particularly susceptible to the risks of flooding and landslides. The rise in soil sealing and the lag in development of urban infrastructure also contribute to heightening the flood risk. The metropolitan areas of Tangiers, Casablanca and Agadir have seen the population of high-risk areas increase significantly over recent decades. With nearly 600,000 inhabitants, the city of Agadir has a population ten times bigger than it did in 1960, when the earthquake decimated around a quarter of its population.

Table 1.5. Population in Morocco (in millions) based on various fertility scenarios between 2004 and 2030

Year	Scenario 1: TFR (Total Fertility Rate) = 1.2	Scenario 2: TFR = 1.5	Scenario 3: TFR = 1.8	Scenario 4: TFR = 2.1	Scenario 5: TFR = 2.47
2004	29.70	29.70	29.70	29.70	29.70
2009	32.00	32.00	32.10	32.10	32.20
2014	34.10	34.30	34.50	34.70	34.90
2019	36.00	36.40	36.80	37.10	37.60
2024	37.40	38.10	38.80	39.40	40.30
2030	38.40	39.60	40.80	41.90	43.40

Source: HCP (2006), Prospects for Morocco 2030: the demographic outlook.

People and strategic sites are also concentrated along Morocco's coastline, an area crucial to the country's economic development. Although the probability of a destructive tsunami remains low, considering this type of risk remains important given the damage it could potentially cause. In 2010, people living on the coast constituted 55% of Morocco's total population. This is where the country's largest cities (Casablanca, Rabat-Salé, Tangiers, Tétouan, Kenitra, Mohammedia, Safi and Agadir) are located and they continue to attract people. In particular, the near-230 km strip between Kenitra and El Jadida is home to 67.5% of the Atlantic coast's urban population. 77% of Morocco's industrial plants and 80% of its industrial jobs are also located on the coastline (Mellas, 2012). Morocco's ports are also situated there, and their importance stems from the fact that almost all of Morocco's foreign trade 98%, according to the Ministry of Public Works, uses maritime transportation routes.

Social vulnerability to natural disasters

From a social perspective, rural poverty and widespread informal housing in urban environments constitute major vulnerability factors for populations exposed to natural hazards. The poorest communities are extremely exposed, as they tend to live in high-risk areas, sometimes in informal settlements, and lack the necessary financial resources to recover from a disaster.

People whose income essentially depends on agriculture-based activities are particularly vulnerable to the risk of drought. When such a risk persists for several years, or recurs regularly, rural households may decide to leave as a way of adapting. They may subsequently contribute, among other factors, to the development of informal housing in metropolitan areas, which is often located in high-risk areas. Consequently, a number of settlements have sprung up on *oued* Bouskoura bed in Casablanca in a context of several waves of migration from rural areas, as well as in the Koulima district of Tétouan that is located in a low-lying area regularly affected by flooding.

The vulnerability of traditional rammed-earth habitats such as *kasbahs* and *ksour* and of ancient centres known as *medina* quarters, which are in particularly poor conditions according to the Ministry of Housing, also heightens social vulnerability. Indeed, the medinas are often home to poor and more vulnerable communities. It is also true for kasbahs and ksour, which have significantly deteriorated over the years and have a low resistance to climate-related risks or risks of geological origin. Access to medinas can also prove difficult in the event of a catastrophe, as their density may result in concentrated damage and complicate emergency services following an earthquake. The isolated nature of some villages and a lack of infrastructure can also hinder rescue attempts, as was the case during the Guelmim floods of 2014, when damage to the town's main access road hampered aid efforts.

Despite the notable progress made in reducing poverty since the 2000s (the number of people below the poverty line dropped from 15% to 9% between 2000 and 2007), Morocco's Human Development Index remains low (130th out of 187 countries in 2013). This is indicated by a high illiteracy rate in rural areas, where substantial pockets of poverty and vulnerability can still be found (AFD, 2014). According to the HCP's *Prospects for Morocco 2030* report, this social vulnerability could be significantly exacerbated by an increase in immigration flows from sub-Saharan regions during the coming decades.

Vulnerability of economic sectors

From an economic viewpoint, aside from the direct disaster damage and losses, certain crucial economic sectors for the Moroccan economy are particularly susceptible to disaster risks. The aforementioned agricultural sector is highly exposed to the risks of drought and flooding. The industrial sector, a growing contributor to the country's economy, is also vulnerable to natural hazards, particularly to floods. Meanwhile, the tourism sector, which represents almost 10% of Morocco's GDP, is especially exposed to the risk of natural disasters.

Vulnerabilities of the agricultural sector

The agricultural sector is of great importance for Morocco's national economy. It represents 15% of GDP and 10% of global product exports, as well as 80% of jobs in rural areas (OECD, AfDB, UNDP 2014). Its high dependency to weather conditions is a major fragility factor for the nation's economy. In 2013, an increase in GDP of 4.7% was largely due to the vitality of this primary sector, while the sector's poor performances in the following year resulted in a drop of GDP back to 2.9%. The agricultural sector has been boosted in recent years by a sector-based plan launched by King Mohammed VI, known as *Plan Maroc Vert* (Green Morocco plan). It aims to encourage investment in the sector, ensure better export crop productivity, and to maintain sustainable family farming, in order to guarantee decent living conditions for rural communities.

Vulnerabilities of the industrial sector

Concerning the industrial sector, an interruption or disruption of manufacturing processes caused by damage to businesses' manufacturing tools or stock during a disaster can have a knock-on effect along the entire production and distribution chains, affecting clients and suppliers alike. This phenomenon is referred to as an indirect or second-order effect, as it is not the result of the disaster itself but of its related consequences. The repercussions can reach areas beyond the ones affected by the disaster, even internationally, and they can linger on well after the event and throughout the rebuilding process.

Morocco's increasing inclusion in global value chains requires a closer look at the issues of business resilience and continuity of services, in the context of the prevalence of stock-less and just-in-time economic models (OECD, 2014). Policies encouraging foreign investment and logistics infrastructure development are at the heart of Morocco's sector-based strategies, which aim to make the most of the country's geographic position at the crossroads of strategic routes for the global economy. The 2008 Tangiers flood caused major damage in the industrial district of Mghogha, which had been built in a flood-risk area. 130 businesses were affected and financial losses were estimated at 1.3bn Dirhams, according to the AZIT, a local industrial association. In addition, Morocco is the world's top exporter of phosphate, holding 28% of global market share (OCP, 2014). OECD peer reviews estimate that disruption to Morocco's supply chains and port infrastructure could potentially affect the global market's phosphate prices, especially in the event of a tsunami.

Improvements to infrastructure and logistics can create opportunities, but they can also increase Morocco's vulnerability to risk if the question of resilience is not sufficiently integrated into design and planning.

Vulnerabilities of the tourism sector

Regarding services, the tourism sector constitutes a significant part of the national economy – 10% of GDP and 7.6% of overall employment (World Travel and Tourism Council, 2014) –; no fewer than 10.18 million tourists visited Morocco in 2015. Tourism was awarded national priority status in the Morocco's Vision 2010 and Vision 2020 plans, which aim to position the country among the world's top tourist destinations (by expanding accommodation capacity, establishing new resorts, etc.), through investments in transport-related infrastructure (airports, high-speed train line between Tangiers and Marrakesh) and hotels.

Disasters do not only affect the economy and tourism infrastructure; they can also directly threaten the tourists themselves, who are generally more vulnerable because of their poor risk knowledge, language barriers, and their reduced vigilance during holiday periods. Furthermore, disasters have an adverse effect on the flow of tourists entering the country. Managing these specific vulnerabilities sometimes calls for joint efforts between Moroccan officials and tourists' countries of origin via diplomatic channels to organise repatriations, as well as cooperation with travel agencies to prevent the arrival of additional tourists in times of crisis. In addition to the loss of short-term tourist revenue caused by trip cancellations, more long-term effects related to Morocco's appeal as a tourist destination can be expected in the event of a sub-standard response from the authorities (Box 1.3).

Box 1.3. Economic impact of disasters on tourism

The impact of disasters on a country's tourism income can be significant: according to the Japanese National Tourism Office, the number of foreign visitors dropped by 50% for several months following the Fukushima Daiichi nuclear disaster, and it took much longer for that figure to return to its pre-disaster level (Iyer, 2012). Similarly, the floods that occurred in Bangkok in 2011 resulted in losses over USD 3bn for the tourism industry (World Bank, 2012), while the three successive hurricanes in the year 2005 and the H1N1 outbreak led to a large reduction in Mexico's tourism revenues (OECD, 2013).

Source: OECD (2013), *OECD Reviews of Risk Management Policies: Mexico 2013: Review of the Mexican National Civil Protection System*, OECD Publishing, Paris, http://dx.doi.org/10.1787/9789264192294-en.

Conclusion and recommendations

Morocco's risk profile reveals a strong level of heterogeneity. The floods that occur on an almost annual basis, often localised but not interrelated, threaten numerous regions of the country and can have major local impacts. Droughts can have a real impact on a significant part of the population and important economic repercussions. The risks of earthquakes and tsunamis, which occur less frequently, involve much greater consequences, both at the local and national level. Other risks must be taken into consideration as part of a multi-risk strategy that would benefit from the sharing and pooling of approaches and resources used to address these various risks, which could result in more effective public policies.

From a macroeconomic perspective, a significant acceleration of growth was noted in Morocco at the beginning of the 2000s. It was propelled by, in particular, an increase in domestic demand and public investment (nearly 5% growth was recorded during the 2000-2009 period). The acceleration of growth has contributed to better well-being for Moroccans, especially in terms of improved access to basic infrastructure (water, roads and electricity), but it has not helped reduce unemployment, which remains high, particularly among young people (nearly 20%) (OCP Policy Centre, 2014). On the other hand, this economic model has reached its limits in light of the current international economic context: while it was exposed to a series of shocks since 2008 that allowed Morocco to showcase formidable resilience. Nevertheless, it also increased the country's deficit and public debt, and revealed structural weaknesses in its development model, which was preventing it from adopting a truly inclusive and sustainable growth approach. This reduced room for manoeuvre will probably force the country to make some slight adjustments in order to continue promoting growth, macroeconomic balance and social stability (AFD, 2014).

In these circumstances, a major shock could have serious economic consequences, considering all of the direct and indirect effects. According to the recent probabilistic analysis carried out by the World Bank (2014), Morocco has a 90% chance of experiencing an event causing losses of 10bn Dirhams within the next 30 years, a figure that represents around 1% of the country's GDP in any given year. Such a shock would potentially have a negative impact on Morocco's growth path and public finances, which have already been weakened by the current economic situation.

The social and political situation has also contributed to make risk management a key public policy issue in Morocco. An effective risk management policy helps to maintain trust in institutions and public authorities, as this is an area of public policy where citizens' expectations are particularly high. While questions of inclusion, transparency and accountability have been at the heart of societal demands in Morocco in recent years, risk management now constitutes a decisive issue.

The significant progress made by Morocco in economic, social, environmental and cultural fields, as well as the country's desire to embark on a long-term development path, are dynamics that encourage the inclusion of critical risk issues in a more assertive manner. Human capital and financial and material resources are more readily available. However, rapid transformations, such as urbanisation and littoralisation, as well as geographical concentration of added economic value, constitute stress factors. If the dominant trends, especially the demographic trends, continue unchanged, vulnerability to critical risks and the impact of adverse events will substantially increase. The issue of critical risks should, therefore, be granted strategic attention in future planning and governance efforts.

However, this risk-focused approach should also be supported by an opportunity-focused approach in order to be truly relevant. In addition to the societal benefits and the peace of mind that an improved integration of the risk dimension can offer, a number of other positive aspects can be identified:

- Risk resilience is featured in development, well-being and competitiveness indicators.
- The country's attractiveness as appreciated by citizens' and investors' partly depends on good risk management.
- The resilience of critical infrastructures and business sectors is contingent upon the quality of prevention and anticipation measures.

- Building risk management expertise can also become an exportable service (with Morocco as a regional service hub).
- A shared risk culture contributes towards societal cohesion and confidence.
- Investing in resilience has high returns. In the long term, the reduction of harmful consequences, a more equal share of risk between stakeholders, and a better integration of innovative technologies all contribute to more sustainable economic development.

Recommendations

PRIORITISE RISK MANAGEMENT IN MOROCCO TO ACHIEVE SUSTAINABLE GROWTH IN THE SHORT, MEDIUM AND LONG TERM

A more holistic approach to risk in Morocco would benefit from a favourable context at this unique moment in Moroccan history. The strong political will, the pursuit of economic growth, the development of major infrastructures, the ongoing decentralisation process and regulatory reforms, the strengthening of sector-based plans, as well as the organisation of the COP 22 meeting in 2016 provide unprecedented opportunities to pursue efforts to incorporate the "risk" dimension into decisions made by public authorities, the private sector and citizens.

Notes

1. The EM-DAT database only takes into account disasters that caused the deaths of at least 10 people, affected at least 100 people, involved a state of emergency or an international aid appeal.
2. Official death toll was 150, with 88 people unaccounted for.

Bibliography

AFD (2011), Cities without Slums, project note, www.afd.fr/home/projets_afd/villes/projets-cld/lutte-contre-exclusion/Morocco-sans-bidonvilles.

AFD (2014), The Moroccan growth model: opportunities and vulnerabilities, http://www.afd.fr/webdav/site/afd/shared/PUBLICATIONS/RECHERCHE/Scientifiques/Macrodev/14-Macrodev.pdf.

Tensift Hydraulic Basin Agency, *Flood risk atlas*, www.eau-tensift.net/fileadmin/user_files/pdf/publications/Atlas_Zones-inondables.pdf.

Balaghi, R., M. Jlibene, B. Tychon, R. Mrabet (2007), "Management of the risk of agricultural drought in Morocco", SECHERESSE, Vol. 18, n°3, July-August-September 2007, http://www.inra.org.ma/environ/docs/articles/balaghi20071.pdf.

African Development Bank (AfDB), Organisation for Economic Cooperation and Development (OECD), United Nations Development Programme (UNDP) (2014), *Economic outlook in Africa 2014.*

World Bank (2014), Building Morocco's resilience: inputs for an integrated risk strategy.

World Bank and Thai Ministry of Finance (2012), "Thai Flood 2011, Rapid Assessment for Resilient Recovery and Reconstruction Planning", World Bank, Washington.

Cherkaoui T E and El Hassani A. Seismicity and seismic hazard in Morocco 1901-2010, Bulletin de l'Institut Scientifique N° 34, Mohammed V University – Agdal, Rabat.

Climate in Morocco, Diagnosis and Prospects, Konrad-Adenauer-Stiftung e.V., Rabat, Morocco.

IAU (2011), North African coastal towns' adaptation to climate change and natural disasters – Phase 1: Risk assessment of the current situation and of the situation by 2030 in Casablanca and Bouregreg Valley – Joint report. *II to the Fifth Assessment Report of the Intergovernmental Panel on Climate Change.*

IPCC (2014), *Climate Change 2014: Impacts, Adaptation, and Vulnerability. Part B: Regional Aspects. Contribution of Working Group.*

IRES, Climate Change Study Programme: State of Play, Presentation by M. M. Messouli on 23 September 2013.

High Commission for Planning (2006), Prospects for Morocco 2030: *the demographic outlook.*

High Commission for Planning (2006), Prospects for Morocco 2030: *Agriculture 2030: What futures for Morocco?*

Hulme M. et al. (2000), Using a climate change scenario generator for vulnerability and adaptation assessment, MAGIC/SCENGEN Workbook.

Iyer, P. (2012), "Now Is the Season for Japan", *New York Times*, 22 March 2012, www.nytimes.com/2012/03/25/travel/a-new-kyoto-opens-its-arms-to-visitors.html?_r=0.

Kaabouben F., Baptista A., Iben Brahim A., El Mouraouah A., Toto A. (2009). On the Moroccan tsunami catalogue, Natural Hazards and Earth System Sciences, n°9.

Ministry of Agriculture and Maritime Fishing (2012), Moroccan agriculture in figures, http://www.agriculture.gov.ma/sites/default/files/agriculture-en-chiffres-2012.pdf, consulted on 1 March 2016.

Water Ministry (2010), National Water Strategy.

Ministry of Public Works, Transportation and Logistics (2014), *Assessment of the exceptional sea swell of 6 January 2014.*

Mokssit, A. 2012, Taking stock of climate change in Morocco, Environment and Changes.

Sahara and Sahel Observatory (OSS), 2008, Moving towards an early warning system for drought in North Africa \ OSS. _ Collection Synthèse n° 4. _, OSS: Tunis, 2008. _ 84 pp., ISBN: 978-9973-856-39-5.

OECD (2011), Future Global Shocks, Improving Risk Governance, OECD Publishing, Paris, http://dx.doi.org/10.1787/9789264114586-en.

OECD (2013), *OECD Reviews of Risk Management Policies: Mexico 2013: Review of the Mexican National Civil Protection System*, OECD Publishing, Paris, DOI: http://dx.doi.org/10.1787/9789264192294-en.

OECD (2014), "A Boost to Resilience: Governing Effective Prevention and Mitigation of Disruptive Shocks", http://dx.doi.org/10.1787/9789264209114-en.

OECD (2014), OECD Study on the Management of Flood Risks: the Seine in Paris Region 2014, OECD Publishing, Paris.

OCP Policy Centre, 2014, Morocco: Growth strategy between now and 2025 in a changing international environment.

OCP (2014), *OCP in figures*, www.ocpgroup.ma/sites/default/files/alldocs/Plaquette_Chiffres_Cles_VF_0.pdf

Samira Mellas. Assessment of the risk of tsunami on Morocco's Atlantic coastline. History. Paul Valéry University – Montpellier III, 2012. French. <NNT: 2012MON30067>, doctoral thesis.

Ministry of Energy, Mining, Water and Environment, Environment department, *Atlas of risks*, http://www.environnement.gov.ma/PDFs/Atlas_Risk.pdf

Ministry of Energy, Mining, Water and Environment, Environment department (2008a), "Mission 1: Identification of risks – risk of flooding", *Study on the mapping of and development of a Geographic Information System for critical risks in Morocco*, Etudes et mesures Les 5 Domaines, www.environnement.gov.ma/PDFs/Rapport_Risque-Inondation.pdf

Ministry of Energy, Mining, Water and Environment, Environment department (2008b), "Mission 1: Identification of risks – risk of tsunami", *Study on the mapping of and development of a Geographic Information System for critical risks in Morocco*, Études et mesures Les 5 Domaines, www.environnement.gov.ma/PDFs/Rapport_Risque-Tsunami.pdf

World Travel and Tourism Council (2014), Travel and tourism: economic impact 2014 Morocco, www.wttc.org/-/media/files/reports/economic%20impact%20research/country%20reports/morocco2014.pdf, consulted on 20 February 2016.

Chapter 2.

Governance of risk management in Morocco

Just like numerous other countries, the strategic framework supporting risk management in Morocco was built gradually, often in the aftermath of major disasters. Despite the considerable advances made in recent years, the current framework is still characterised by a plethora of isolated initiatives. Their impact remains restricted by a lack of coordination in the absence of a platform linking the capabilities and roles of all involved stakeholders. This chapter describes the regulatory and institutional framework for risk governance in Morocco[1]. It will look at the definition of individual roles and responsibilities, and assess the overall consistency of the system and its ability to reach set targets. It examines the effectiveness of existing measures, and proposes – on the basis of international experience – recommendations to improve risk governance in Morocco.

1 Information on the regulatory and institutional framework of risk governance in Morocco is correct and applicable as of 30 October 2015.

Introduction

Morocco is exposed to a rising number of risks, resulting from external factors (climate change, international interdependencies) and internal factors (population growth, urbanisation, growth of critical networks and sectors, economic diversification). It is, therefore, increasingly important for Morocco to establish a governance framework for risk management that can help meet these challenges.

Against a backdrop of ever-increasing socio-economic and environmental uncertainty, enhancing awareness and consideration for critical risks in public policies is essential to ensure better resilience. This essentially horizontal issue raises numerous challenges, particularly for empowering a wide range of stakeholders, aligning priorities and involving stakeholders.

To respond to these challenges, the OECD Recommendation on the Governance of Critical Risks calls for the development of a national risk governance strategy. Setting up a holistic and multi-risk governance structure would help establish national leadership and allocate relevant roles and responsibilities in order to ensure better resilience when confronted with the adverse effects of critical risks. Such a strategy would set objectives for each stage of the risk management cycle, and define priorities for prevention, mitigation, emergency response, recovery and reconstruction. It would also ensure these priorities are incorporated into different sector-based strategies and development plans drawn up by public institutions.

The OECD Recommendation invites governments to establish clear stewardship, coordinate the participation of all stakeholders – at national and local level – and adopt an inclusive dialogue on risk policies, including with civil society and the private sector. This will strengthen the involvement of society as a whole and help define a common vision of critical risks, thereby guaranteeing better results with regard to resilience.

This chapter will firstly describe changes to risk governance in Morocco and highlight the challenges raised by the large number of actors, sector-based plans and initiatives involved. Secondly, it will touch upon risk governance structure options, based on international experience.

Strategic approach to risk management in Morocco

The strategic framework for risk management in Morocco is not yet based on an integrated and cross-cutting policy. As in many countries, it is constituted by the parallel presence of civil defence, land-use, town planning, water, environmental and agriculture policies, among others. This process brings together a number of actors, and involves responsibilities that are shared vertically, across different levels of government, and horizontally, across various sectors affected by the issue of risk (water management, infrastructure management, etc.). To improve risk governance frameworks, a better understanding of how these different public policies have evolved is required.

Changes to risk management policies in Morocco

Similar to other countries, risk management policies in Morocco were initially mainly based on the emergency response to disasters. Large structural preventive measures were favoured at first. The country experienced a learning process following each major disaster and following the implementation of each international agreement. This has reinforced the recognition that all of the elements of the risk management cycle must gradually become integrated, as has been suggested by the OECD Recommendation on the Governance of Critical Risks.

Initial elements of a risk management policy in Morocco in modern times: emergency response and structural measures

Morocco's modern history has been marked, as far as risk management is concerned, by the 1960 Agadir earthquake. Aside from the introduction of the paraseismic "Agadir standards", which were particularly innovative at the time, the disaster instigated the development of a civil defence policy that was coordinated by the Prime Minister's office and was aimed at developing the country's disaster-response capabilities. In parallel, the objective of the large dams policy launched by King Hassan II was to develop Morocco's agricultural potential, as it also helped prevent risks of drought and flooding. Moreover, the major droughts at the beginning of the 1980s accelerated this public policy. In 1995, the devastating Ourika Valley flood cast a shadow over the country leading to the 1997 reforms of civil defence policies, and the establishment of regional commands to react more quickly to local disasters.

Dynamic prevention-based approach in the 2000s and the response to the Al Hoceima earthquake

Since 1999, Morocco's public policies have been redeployed. The development of sector-based policies and an increased focus on the issue of sustainable development have enabled the enhancement of risk management policies in many ways, particularly in the area of prevention. Line ministries became involved: the Ministry of Housing put together the national paraseismic regulations that were approved in 2004, the Ministry of Water launched the National Flood Protection Programme in 2003 (following the catastrophic Mohammedia flood in 2002), and the Ministry of the Environment encouraged a national dialogue on risks in 2003as part of its National Action Plan for Environmental Protection. The latter led to the creation of the first national risk atlas in 2008, developed in partnership with Morocco's entire scientific and technical institutions, whose resources and capabilities were simultaneously bolstered. In addition, the *Maroc Vert* (Green Morocco) plan, launched by the King Mohammed VI in 2008 and implemented by the Ministry of Agriculture, includes a significant component dedicated to the fight against drought.

The 2004 Al Hoceima earthquake constitutes the third worst disaster in Morocco's modern history. The human and material loss from the catastrophe prompted serious questions on the country's disaster-response capabilities and culminated in the establishment of a Royal Commission. This, in turn, led to the creation of a structure tasked with coordinating all emergency management actors under the umbrella of the Ministry of the Interior: the Monitoring and Coordination Centre (CVC). The CVC is now the first pillar of Morocco's integrated risk management approach. In parallel, other sector-based prevention efforts have also increased since the Al Hoceima earthquake.

Moving towards an integrated risk management approach in Morocco?

In recent years, Morocco has sought to adopt a more integrated risk management approach, particularly with support from international partners. Among other improvements, this has resulted in the promotion of a financial instrument that is specific to risk management, known as the Fund to Combat the Effects of Natural Disasters.

Since 2013, the Ministry of General Affairs and Governance has been responsible for developing a national risk management strategy.

Legal and regulatory framework for risk management

This historical perspective on the evolution of policies and instruments related to principal risk management reveals a certain dynamic. While Morocco does not currently possess a national risk management strategy, it has gradually been putting in place a legal and regulatory framework covering the main components of the risk management cycle, but not all of it. This framework is characterised by a plethora of laws and legislations (Table 2.1), and a predominant single-risk approach.

Table 2.1. Main laws and regulations related to risk management in Morocco

Year	Legislation	Topics				
		Strategic coordination	Assessment	Prevention	Emergency management	Rebuilding/ Recovery
1955	Dahir (King's decree) of 30 April1955 on civil defence	•			•	
1977	Dahir enacting Law n° 1-75-168 of 15 February1977 on the responsibilities of governors	•			•	
1995	Law n° 10-95 on water			•		
1997	Decree n° 2-97-176 of 15 December 1997 on the remit and organisation of the Ministry of the Interior			•	•	
2003	Law n° 11-03 on the protection and improvement of the environment			•		
2003	Law n° 12-03 on environmental impact studies		•			
2004	Decree n° 2-04-267 of10 May 2004 approving regulations related to earthquake-resistant construction (RPS 2000)			•		
2009	Financial Law n°40-08 corresponding to the 2009 financial year, establishing the Fund to Combat the Effects of Natural Disasters			•		•
2013	Decree n° 2-12-682 of 28 May 2013 modifying Decree n° 2-02-177 of 22 February 2002 approving regulations related to earthquake-resistant construction (RPS 2000, version 2011)			•		
2013	Decree n° 2.13.253 of 20 June 2013 and Decree n° 2.13.836 of 13 November 2013 on the remit of the Ministry of General Affairs and Governance	•				
2014	Framework Law n° 99-12 on the national environmental and sustainable development charter			•		
2015	Organic Law n° 111-14 on regions			•		
2015	Organic Law n° 112-14 on préfectures and provinces			•		
2015	Organic Law n°1 13-14 on municipalities			•	•	
2016	Law establishing the insurance scheme for consequences of catastrophic events, and modifying and completing Law n° 17-99 on insurance codes (Draft)					•

Note: Please see the detailed description of laws and regulations in Appendix F, which also includes laws on health, education and critical infrastructure.

Source: OECD

A structured framework for emergency response

Morocco's emergency response policy is the most advanced example of an integrated multi-risk approach based on a well-established framework, and placed under the authority of the Ministry of the Interior. The *Dahir* of 30 April 1955 on civil defence, as well as Decree n° 2-97-176 of 15 December 1997 on the remit and organisation of the Ministry of the Interior, entrusted the Directorate-General of Civil Protection (DGPC) with organising emergency services during disasters: "the goal of civil defence, in times of peace, is to implement and coordinate emergency services in the event of major disasters" (Article 1). The regular revision of legislation determining the organisation of civil defence through a variety of memoranda and/or decrees also demonstrates the flexibility of such a framework.

Emergency response is managed at national level by the CVC, and at regional level by the *walis* and governors of *préfectures* and *provinces*, adhering to the principle of subsidiarity. The *Dahir* enacting the Law of 15 February 1977 on the responsibilities of governors includes a provision on "the power of interministerial coordination" and provides a legal basis for risk-related actions.

Dispersed prevention policies

Contrary to emergency management policies, prevention policies are not supported by a well-established framework covering all possible issues. Indeed, sector-based and single-risk approaches that correspond to structural provisions with top-down tactics continue to prevail. In addition to prevention-related provisions incorporated into civil defence legislation, these sector-based approaches include laws and regulations that relate to water management, environmental protection, housing rules, and the responsibilities of municipalities.

Thus, flood-risk prevention is based on the 1995 Law on water. Introducing the concept of basin management, the law led to the creation of Hydraulic Basin Agencies (ABH) for Morocco's nine major river basins. The law defines the responsibilities of these agencies, charged with building the necessary infrastructure to prevent and combat flooding. It includes explicit provisions regarding the risk of flood, as well as the prohibition of construction without prior approval, which would obstruct the flow of water. The law also establishes complementary prevention provisions to combat flooding within the framework of the management of the Public Water Domains. The ABHs are, therefore, responsible for drafting and ensuring the implementation of the Master Plan for the Integrated Management of Water Resources (PDAIRE), whose guidelines are systematically integrated into urban planning documents, suggesting and executing regulatory measures in cases of a declared water shortage or to reduce the risk of flood,, and to build infrastructure that will help prevent flooding.

As regards the risk of earthquakes, Morocco possesses paraseismic regulations, known as the RPS 2000, established by Decree n°2-02-177, which also established the National Paraseismic Engineering Committee. The RPS aims to limit buildings' vulnerability to seismic tremors by setting construction rules for the country's major seismic zones. These regulations were revised in 2008 under the auspices of the Ministry of Housing.

The risk of tsunamis, or more broadly, the risk of marine submersion, does not yet have any specific prevention framework.

The legislative and regulatory corpus related to the environment has undergone significant development in recent years, with increased awareness of sustainable development issues, which partly include risk prevention. Law n° 11-03 on the protection and development of the environment, promulgated in 2003, allows for the implementation of protective measures to fight against desertification, flooding, deforestation, erosion, the loss of cultivable land, and soil pollution and its natural resources, although these measures have not yet been enforced. In the same year of 2003, the promulgation of Law n° 12-03 on environmental impact studies made it compulsory to carry out a prior assessment of any environment-related project that involves activity, work, construction or structural work and its possible repercussions by any natural or legal person. This law gives a significant control power to local commissions attached to the Ministry of Environment to account for the issue of risks and flooding, in particular, in construction work. Finally, Framework Law n° 99-12 on the national environmental and sustainable development charter, enacted in 2014, includes the development of a legislative framework related to risk prevention. This particular attention given to environmental issues could offer an opportunity for improved harmonisation of prevention policies when developing laws to implement them.

Finally, from a local standpoint, the new Organic Law on municipalities, especially Articles 85, 100 and 101, stresses the importance of local authorities in the prevention process. These articles assign to municipalities the responsibility of ensuring the enforcement of town planning laws and regulations, and adherence to prescriptions in development plans and town planning documentation, issuing building permits, and to take any necessary measures to prevent flooding and disasters. In Morocco, town planning legislation partially integrates the issue of risk, but its implementation is hampered by the lack of a unified risk mapping. Following the Al-Hoceima earthquake, a draft law – Law 04-04 – was drawn up to transfer certain competencies from local elected officials to the state, and to reinforce penalties for the issuance of building permits in high-risk areas. However, the parliamentary debates did not culminate in an agreement that could lead to a vote on these provisions.

This area of town planning legislation could be opportunely expanded to include the issue of technological and industrial risks. A *Dahir* enacting regulations for unhealthy, inconvenient or dangerous establishments, issued on 25 August 1914, remains the main basis of legislation on potentially dangerous industrial facilities. Several *Dahirs* since then, such as the ones in 1933 and 1950, have clarified the scope of the 1914 Law. There are plans to highlight technological risk prevention plans in town planning documentation and raise residents' general awareness, but the regulations have not been formalised.

Draft law on compensation for disaster damage

The Ministry of Economy and Finance recently created a draft law (see Chapter 6) setting up insurance coverage for the consequences of natural disasters. This aims to guarantee a minimum right to compensation for bodily harm or loss of the use of a principal residence during a disaster through a compensation mechanism for victims of disasters. It combines an insurance system for individuals in possession of an insurance policy and an allocation system for individuals with no coverage. The draft law was adopted by the Governing Council in March 2016, and its approval by the Parliament is scheduled for 2016, with implementation set for 2017 to plug a gap in the risk management cycle.

Shortcomings of the governance framework for risk management

The notion of risk management is by nature cross-cutting and horizontal. Consequently, the establishment of a dedicated legal and regulatory framework that is consistent and able to react throughout the entire risk management cycle is often necessary to clearly establish roles and responsibilities. According to the experts and relevant stakeholders, the framework in place in Morocco, as described in this section, demonstrates a lack of consistency. Several actors believe the legal framework requires a review. The issues of risk assessment, town planning or compensation are not sufficiently covered thus far. A clarification of responsibilities would help empower key stakeholders, limit inconsistencies, and over time, improve resilience. Given the cross-sector nature of these public polices, the question of their coordination is a crucial one. This observation led the Ministry of General Affairs and Governance to become concerned with this issue. In fact, it was officially given the responsibility of developing a national strategy to confront risks by the Decree n° 2.13.253 of 20 June 2013.

Laws and regulations also help introduce control and sanction procedures, which can be useful risk management tools (e.g. in urban planning), but not exclusively in that domain. However, other public policy tools are also available to pursue effective risk management policies. Legal congestion and the difficulty of enacting laws on these subjects hamper progress on risk-related public policies. The draft Law on town planning, which includes specific provisions related to the integration of risk management into urban planning, and the draft law on the establishment of a compensation system in the wake of disasters have been subject to discussion for many years. The complicated nature of this type of legal "journey" demonstrates lawmakers' low-priority of issues relating to risk management.

Risk management in sector-based plans and programmes in Morocco

However, the regulatory framework's shortcomings are not preventing sector-based action. Since the start of the 2000s, Morocco has, to a great extent, favoured the development of sector-based strategies to support its economic and social development. This approach enables long-term visions (from 10 to 30 years in general) and planning, as well as forward-looking objective-setting. It is also particularly appropriate for risk management, which requires a long-term perspective and the inclusion of strategic forecasting elements. Despite the development of numerous medium- to long-term action strategies, plans and programmes over the past 15 years (Table 2.2), this dynamic did not specifically concern risk management, which consequently could not benefit from the development of an integrated strategy. The cross-sector nature of risk management and the plethora of stakeholders involved explains why this was not the case, given the majority of these strategies are sector-specific.

Table 2.2. Principal development plans and sector-based strategies

Sector-based strategy/plan	To be completed by	Main actor	Objective
Sector-based plans related to risk management			
National Flood Protection Plan	2017	Ministry of Water	Mobilise resources for preventive measures at 391 sites at risk of flooding
Green Morocco Plan	2020	Ministry of Agriculture and Fisheries	Encourage balanced and sustainable development of the agricultural sector in Morocco & highlight the agricultural potential across the country
National Water Strategy	-	Ministry of Water	Meet Morocco's long-term water needs and deal with the effects of climate change through coordinated management of supply and demand and by preserving resources
National Water Plan	2030	Ministry of Water	Determine national priorities and draft an action plan related to the mobilisation and use of water resources
Master Plan for the Fight against Forest Fires	-	HCEFLD	Prevent forest fires through prevention, detection, early warning and emergency response measures
National River Basin Development Plan	2016	HCEFLD	Fight against water erosion, the silting-up of dams, and flooding through the conservation of natural resources in river basins
Related sector-based programmes			
National Sustainable Development Strategy	2020	Ministry of Environment	Implement a green and inclusive economy by raising awareness of environmental problems
National Plan of Action for Desertification Control	-	HCEFLD	Mitigate the effects of drought and desertification by dealing with socio-economic factors in particular.
National Territorial Development Plan	2027	Ministry of Urban Development and Regional Planning	Define the future organisation of the regions in a way that ensures sustainable development, social equality and economic efficiency
Social strategies			
National Sewage Treatment and Wastewater Purification Programme	2020, 2030	Ministry of Environment	Achieve an overall urban connection rate to the sewage system of 75% in 2016, 80% in 2020 and 100% in 2030; achieve a volume of treated water of 50% in 2016, 60% in 2020 and 100% in 2030
National Human Development Initiative	-	On the initiative of King Mohammed VI, this is now run by a collection of committees at central, regional, provincial and local level	Reduce poverty, instability and social exclusion, by improving access to and quality of basic services and infrastructure, and providing support to vulnerable communities and income-generating activities
Cities Without Slums Programme	2012	Ministry of Housing and Urban Policy	Definitively eliminate shanty towns in urban environments (concerns 1.6 million inhabitants in nearly 85 towns)

Table 2.2. Principal development plans and sector-based strategies *(continued)*

Sector-based strategy/plan	To be completed by	Main actor	Objective
Sector-based economic strategies			
Maroc Export Plus	2018	MCINET	Triple exports of goods and services in ten years, excluding phosphates and derivatives
Industrial Acceleration Plan	2020	MCINET	Increase the industrial share in the nation's GDP by 9 points and create 500,000 jobs, adopting a new approach based on the setting-up of well-functioning ecosystems
Halieutis Plan	2020	Ministry of Agriculture and Fisheries	Triple the fisheries' share in national GDP and sustain an increase in the consumption of fish
Tourism Vision 2020	2020	Ministry of Tourism	Double tourism's share in national GDP and make Morocco one of the top 20 holiday destinations around the world, and a reference for sustainable development in the Mediterranean Region
Energy Strategy	2030	Ministry of Energy and Mining	Ensure the sustainability of primary energy supply, develop electrical production capacity, promote energy-saving measures and renewable energy
National Logistic Competitiveness Development Strategy	2030	METL	Strengthen logistical competitiveness to improve Morocco's position as a trading partner
National Ports Strategy	2030	METL	Promote the ports' performance as a necessary element in positioning Morocco as a logistics platform in the Mediterranean Basin.

Source: Interviews conducted by the OECD.

However, Morocco's sector-based strategies have had some impact on risk management. In the absence of an overall strategic framework, risk management policy in the country developed partly within sector-based plans and programmes that were put in place by various national institutions. Supported by individual actors, these plans and programmes often favoured a single-risk approach, without harnessing the synergies offered by an integrated risk management approach.

- Regarding flooding, the National Flood Protection Plan (PNPI), launched in 2002 by the Ministry of Water, identified 391 nationwide endangered sites, including 50 priority locations, and mobilised resources to introduce risk reduction measures, with the aim to mitigate flood risk in all the sites by the year 2017. By 2015, 244 sites had been reviewed, including all priority sites, and a new programme is currently under development. The PNPI is fully integrated into the National Water Strategy (SNE) that was launched in 2009.

- Drought risk management, which is not subject to a specific strategy, is incorporated into the two major agriculture and water strategic plans. The Maroc Vert (Green Morocco) Plan, launched by King Mohammed VI in 2008, aims to increase agricultural GDP by 2020, while strengthening resilience to the risk of drought through prevention programmes such as (i) the National Irrigation Water Saving Programme (PNEEI), which features the development of water-saving irrigation techniques and a progression towards high added value crops, (ii) the conversion of 1 million hectares of grain-growing farmland in at-risk areas into fruit-growing zones that are more resistant to drought, and (iii) the development of agricultural insurance. In parallel, the National Water Strategy, adopted in 2009, also includes the continued pursuit of a water supply policy by scheduling major infrastructure work (dams and water transfer) and unconventional uses of water (wastewater purification and sea-water desalination). The National Water Strategy pays special attention to the development of drought management plans for river basins.

- An interesting case is the Master Plan for the Fight against Forest Fires, adopted in 2001, as it is the most accomplished example of a risk-specific, integrated sector-based strategy that has achieved tangible results. Under the guidance of the High Commission for Waterways and Forests and Fight against Desertification, this plan integrates all actions related to the risk of forest fires. It is supported by a clear legislative and institutional framework and sustainably financed. The plan has achieved notable results, since its inception in 2007. The total area affected by fires was significantly reduced, dropping from 14 hectares per fire during the 1961-1970 period to just 7 per fire during the 2006-2015 period.

Related sector-based programmes

Aside from strategies related to the water and agriculture sectors, a number of other sector-based plans are linked to the issue of risk. These include numerous action plans related to the environment, sustainable development and climate change, regional development programmes, certain social programmes, and the economic development strategies of different sectors. Some incorporate risk directly, but the majority do not, or only partially and with no subsequent effect. This is a major impediment raising awareness and consideration of risk management in Morocco's socio-economic development framework.

The majority of environment-related programmes touch on the issue of natural risks, and weather-based risks, particularly flooding and drought. Thus, one of the objectives of the National Sustainable Development Strategy adopted in 2015, is the creation of a national climate risk prevention and response plan, in conjunction with the National Climate Change Policy, adopted in 2014. This is also the case for the National Plan of Action for Desertification Control (PAN-LCD), updated in 2012, which incorporates sustainable management and land conservation to address the risks of drought and flood, or the National River Basin Development Plan, which contributes towards flood-risk management.

Regarding regional development and town planning, which also remain major challenges for risk prevention in other OECD countries, the issue of risk is clearly lacking. In 2004 the National Territorial Development Plan (SNAT) helped build momentum about regional and infrastructure development, but did not explicitly address the question of risk. In the absence of country-wide risk mapping, the Master Plans for Urban Development (SDAU) do not systematically include the problem of risk from a local perspective, and neither do the Community Action Plans (formerly Communal Development Plans) initiated by regional authorities.

Social plans and strategies, or their updated versions, do not specifically include the issue of risk in their objectives either, although they can sometimes be mobilised to finance certain related elements. The Cities without Slums Programme contributed to reducing the exposure to flood risk among vulnerable neighbourhoods. The National Liquid Sanitation Programme financed the creation of numerous storm water drainage systems, thereby playing a key role in flood prevention. Projects attached to the National Human Development Initiative have also sometimes helped to reduce the vulnerability of poor or isolated communities by improving their level of resilience.

Similarly, the principal economic development strategies, particularly the Industrial Acceleration Plan or the National Logistic Competitiveness Development Strategy, do not mention risks (natural or industrial) as potential threats or opportunities. The Tourism Vision 2020 strategy, meanwhile, includes the notion of sustainability and accounts for the environmental and ecological constraints in regions that are popular for tourists, but it does not allow for risk mitigation measures. These strategies have a significant impact on the medium to long-term development of the Morocco's economic and industrial potential, particularly in terms of establishing manufacturing plants, and they offer opportunities to directly integrate resilience into investment decisions. While these major sector-based strategies have boosted the country's socio-economic development for around a decade, it would appear pertinent to better integrate the risk dimension in a way that transparently provides the best possible visibility of opportunities and risks.

The limited integration of risk management in sector-based policies and national development plans raises fundamental questions about transparency. A greater consideration of risk management in high-level strategic documents that serve as a reference for several public policies in Morocco should lead to) increased conformity of sector-based strategies.

Multiple risk management players in Morocco

The issue of governance is especially pertinent for the design of a holistic, strategic framework for risk management in Morocco given the multiplicity of actors involved and their often uneven contributions. As in many OECD countries, risk management in Morocco requires the interaction of a wide range of public authorities at the central and local levels, the scientific community and, although to a lesser extent, civil society and the private sector (Table 2.3). This represents several governance challenges, particularly for the coordination of policy definition and implementation. To better identify these challenges, a second step consists of clearly understanding the respective roles and responsibilities of the different actors.

Principal central government actors

Although the 2011 Moroccan Constitution initiated a movement towards greater decentralisation, Morocco's national government remains very centralised. Government institutions and their local representatives are at the forefront of the definition and implementation of public policies on risk management.

The **Ministry of the Interior** plays a central role in risk management in Morocco. It is the primary actor in emergency management, defines and implements policies through the DGPC and coordinates activities at the national interministerial level through the Monitoring and Coordination Centre, and at the local level through the regional *walis* and prefectural and provincial governors, who are responsible for government activities locally. With its Territorial Communities Department, the Ministry is also involved in the entire risk management cycle regionally, as the main intermediary between the government and regional authorities, particularly on issues relating to technical and financial assistance. Finally, the Ministry of the Interior administers the Fund to Combat the Effects of Natural Disasters and, as a result, is involved in coordinating preventive actions financed by this fund.

The **Head of Government's Ministry of General Affairs and Governance** contributes to the development and monitoring of government policies on governance. This Ministry has made risk management one of its top priority areas. New governance mechanisms should be developed to ensure a cross-cutting approach to risk management. Since 2013, the Ministry of General Affairs and Governance has been a leader in developing an integrated risk management strategy at the national level with other stakeholders.

Table 2.3. Principal Risk Management Stakeholders in Morocco

	Coordination	Assessment	Prevention	Emergency	Reconstruction
Government bodies					
Ministry of the Interior	●		●	●	
Territorial Communities Department			●		
Directorate-General of Civil Defence				●	
Monitoring and Coordination Centre	●				
Security services (police, gendarmerie, etc.)	●			●	
Ministry of General Affairs and Governance	●	●			
Ministry of Economy and Finance		●			●
Ministry of Energy, Mines, Water and Environment		●	●		
National Meteorological Department		●		●	
Hydraulic Basin Agencies		●	●		
National Office of Electricity and Potable Water		●	●	●	
Ministry of Equipment, Transport and Logistics		●	●	●	●
National Ports Authority			●	●	
National Airports Authority			●		
Morocco Roads			●	●	
National Railways Authority			●		
Ministry of Housing and Urban Policy			●		●
Al Omrane Group			●		●
Ministry of Urban Development and Regional Planning			●		
Urban Agencies			●		
Ministry of Agriculture and Maritime Fishing		●	●		
Ministry of Health				●	
Ministry of National Education and Professional Training			●		
High Commission for Water and Forests and the Fight Against Desertification		●	●	●	

Table 2.3. Principal Risk Management Stakeholders in Morocco *(continued)*

	Coordination	Assessment	Prevention	Emergency	Reconstruction
Territorial Communities					
Municipalities			●	●	●
Provinces and prefectures			●	●	
Regions			●	●	
Private Sector and Insurance Companies					
CGEM			●		
Maroc Telecom			●		
Bank Al Maghrib			●		
Subcontractors for electricity services, water and liquid sanitation			●	●	
Insurers					●
Civil Society					
Civil society				●	●
Non-governmental organisations			●		●
Red Crescent			●	●	
Research Institutions					
Universities		●			
National Centre for Scientific and Technical Research		●			
Royal Centre for Remote Sensing		●			
Rabat Scientific Institute		●			
Royal Institute for Strategic Studies		●			
Hassan II Agronomic and Veterinary Institute		●			

Source: OECD, based on interviews

The **Ministry of Agriculture and Maritime Fishing** is at the forefront of addressing the risk of drought and includes the protecting plants and animals at the national level among its missions. Within its Green Morocco Plan, the Ministry has a wide-range of technical and financial tools for managing drought risk. These include incentives aimed at cutting the risk *ex ante* by reducing the exposure of Morocco's agriculture to drought and *ex post* measures to reduce damage, and at developing insurance mechanisms. It has a large decentralised network across the country composed of regional agricultural development offices, and regional and provincial agriculture departments.

The **High Commission for Water and Forests and the Fight Against Desertification** (HCEFLD) develops and implements government policies on conservation and sustainable development of forest resources and the fight against desertification. HCEFLD is particularly active in managing forest fires in Morocco. In partnership with various other agencies, including the DGPC, the Royal Gendarmerie, the

Royal Armed Forces, the Auxiliary Forces, local authorities, the Ministry of Equipment and Transport, and the National Airports Authority, it conducts an annual prevention programme and forest fire management, combining prevention and awareness-raising measures, detection and warning systems, and it implements emergency response plans. It also participates in flood risk mitigation through river basin management.

The **Ministry of Economy and Finance**, through its Insurance and Social Contingencies Department (DAPS), works towards improving the country's natural disaster insurance system. Since Morocco has a low rate of insurance coverage and its traditional insurance policies do not include disaster risk coverage, the Ministry of Economy and Finance is seeking to establish a new coverage system.. The Government Council adopted a draft law in March 2016, with Parliamentary approval planned in 2016 and implementation beginning in 2017.

The **Ministry of Water** is responsible for long-term flood prevention policy and water resource management, in order to ensure a balance between supply and demand, and for short-term water shortages (occurring as a result of drought, natural disasters or other causes). The Ministry develops specific public policy tools (National Water Strategy, National Flood Protection Plan) that are financed and implemented regionally. This is accomplished through the Hydraulic Basin Agency system, which enjoys both regulatory and monitoring authority. The Ministry also relies upon the National Meteorological Department, which oversees the activities related to meteorological and climatological observation and forecasting, playing a key role in the early warning system for hydrometeorological risks.

The **Ministry of Environment** develops and implements government policies relating to the environment and sustainable development. As the focal point for the United Nations International Strategy for Disaster Reduction, and permanent member of the European and Mediterranean Agreement on Major Risk (Eur-OPA), the Ministry started taking preventive measures in Morocco in 1995. As a guarantor for the implementation of the International Charter for Sustainable Development, it is responsible for ensuring the integration of sustainable development principles with the country's public policies. It also plays a coordinating role in the definition of Morocco's climate policy. It provides follow-up reviews based on environmental impact studies in different regions.

The **Ministry of Public Works, Transport and Logistics** plays a key role in the resilience of Morocco's critical infrastructure. It develops and implements government policy relevant for roads, ports, railways, air and maritime issues. This role has gained further importance after Morocco made infrastructure development a top priority in an era of increasing globalisation. The Ministry oversees strategic agencies, such as the National Ports Authority, the National Airports Authority, and the National Railways Authority. It has undertaken an ambitious risk management programme with these agencies that includes floods, storm surge, snow and sandstorms, It emphasised emergency management and established crisis units at the central and regional levels.

The **Ministry of Housing and Urban Policy** is an essential stakeholder for Morocco's housing policies. It participates in the development of and follow-up on construction regulations, particularly the RPS 2000 regulation on seismic construction standards, which was revised in 2011. With the support of METL, the Ministry initiated the Ground Construction Paraseismic Rule, RCPT 2011. The Ministry is also responsible for the City Without Slums Programme, which aims to eradicate shanty towns in Morocco, and oversees the Al Omrane Group, a public housing construction company that aims to improve the availability of dwellings in Morocco and assist cities in the development of large-scale integrated urban projects that balance urban, social, economic and environmental planning. The Ministry and Al Omrane could be required to contribute to the allocation of housing for victims of natural disasters.

The **Ministry of Urban Development and Regional Planning** is responsible for developing and executing government policy for regional development, urban development and architecture. It prepares key urban planning documents, such as the National Regional Development Plan (SNAT), the Regional Development Plans (SRAT), and the Master Plan for Urban Development (SDAU). At the local level, the Ministry oversees urban agencies, which in turn are responsible for designing land use plans such as Urban and Rural Development Plans, in accordance with SNAT, SRAT and SDAU. Urban agencies also provide guidance on urban development projects, which gives them an important role in risk management.

Other ministries, such as the Ministry of Health and the Ministry of Education, are also key participants in the risk management process. The **Ministry of Health** plays an important role in emergency management, through its *National Medical Emergency and Health Risk Management Strategy* for disasters, developed in 2005 with the support of the World Health Organisation. This strategy mandates the development of hospital emergency plans for health emergencies or disasters. The Ministry also participates in assessment studies, and has developed a map of epidemiological risks across the country. The **Ministry of Education** ensures school safety and is responsible for the development of a risk culture among students. The **Ministry of Higher Education and Scientific Research** supports research efforts in the area of natural risks and provides accreditations allowing Moroccan universities to provide training on natural risks.

Local approach

Risk management policy also involves vertically shared responsibility.

Local authorities

At the institutional level, Morocco is organised into 12 regions, 75 *préfectures* or *provinces* and a total of 1,503 municipalities. Various regional authorities are represented by elected councils and have statutory responsibilities for regional and economic development, as well as for some public services. The Ministry of the Interior allocates budgetary resources to these regional authorities. In accordance with the requirements of the new Constitution established on 29 July 2011, a process of regional reorganisation is currently underway in Morocco. Decentralisation laws approved in 2015 established a new regional division and clarified different responsibilities. The division of regions, which were reduced from 16 to 12, is set to provide a structure for assigning special areas of responsibility.

Morocco's 1,503 municipalities play an essential role in risk prevention, based on the prerogatives defined by Organic Law 113-14 on municipalities (see below). They are responsible for the organisation, coordination and follow-up of local services provided to their residents. Concerning risks, they monitor the requirements of urban planning documents and grant building permits. They participate in the development of activities and small and medium-sized water projects for rainwater management and flood protection. They also contribute to environmental protection, coastline management, the development of parks, cliffs, lakes and riverbanks, the management of national roads, and the improvement of medinas. Municipalities take the necessary measures to prevent fires, disasters, floods and other public catastrophes. They may also outsource some public services, particularly water and sewer management, to outside companies, which makes them important actors for the resilience of critical infrastructure. Since 2009, municipalities have been required to develop a Community Action Plan (PAC) (formerly known as Community Development Plan) that lists comprehensive action plans for the upcoming six years aiming for the sustainable development of the municipality's territory. Given that PACs are developed through the use of a participative approach, they may include problems related to risk management, although this is relatively recent.

Regions, *préfectures* and *provinces* are responsible for regional development, environmental protection and water management. In accordance with Organic Laws 111-14 on the regions and 112-14 on the *préfectures* and *provinces*, regions are responsible for the economic, social, cultural and environmental development, as well as for regional development. Meanwhile, *préfectures* and *provinces* are responsible for encouraging social development. The implementation of the presently proposed regionalisation system will require elected Regional Councils to implement regional development plans.

Decentralised government agencies

Decentralised government administration is under the authority of the regional *wali* or provincial governor, who may also be locally represented in rural circles and communities, as well as in urban districts.

Wilayas are responsible for carrying out government activities at the regional level. They provide supervision over local elected assemblies and coordinate the activities of various departments through local representatives. The *wali* is in charge of maintaining public order, applying regulatory frameworks and implementing public policy to execute the government directives, as well as assisting local communities. The *wali* plays an important role in coordinating all crisis management activities. His decision-making powers are significant. He can allocate resources and prioritise actions from a command post he directs in the event of a crisis, and he is able to convene several different departments. The *wali* is also a key intermediary between the central government and different local authorities, and plays a major role in liaison of crisis communications and information.

In *préfectures* (primarily urban areas) and in *provinces* (primarily rural areas), the Governor monitors the application of *Dahirs*, laws, and regulations, as well as the execution of government decisions and directives. The Governor is also responsible for maintaining order. He coordinates the activities of external government agencies, public entities whose activities do not cover areas beyond the *préfecture* or *province*.

Various departments also have regional representatives that are essential actors in the implementation of risk management policies on the ground. Urban agencies are responsible for developing urban planning documents, Urban Master Plans (SDAUs) and Development Plans (PAs), and for monitoring their implementation. The Hydraulic Basin Agencies implement water policy thoughout Morocco's nine major river basins. They develop and implement Draft Master Plans for the Integrated Management of Water Resources in each river basin, particularly related to floods. They also monitor the Public Water Domain. The Hydraulic Basin Agencies are responsible for managing Morocco's principal dams. They establish management plans for domestic and industrial water based on the usage type set by a local office of the National Water and Electricity Authority, or the Ministry of Agriculture for irrigation water.

The Moroccan government has also established agencies dedicated to financing and implementing development projects in specific regions. These include the Agency for the Promotion and Development of the North, a dedicated space to coordinate the emergency programme that followed the 2004 Al Hoceima earthquake. Similarly, the Bouregreg Valley Development Agency, the National Agency for the Development of Oasis Areas and the Argan Plantation and the Eastern Agency also direct structural projects that can incorporate risk management issues, particularly those related to the prevention of critical risks.

Challenges for multi-player coordination

Towards improved national coordination

In recent years, there has been progress towards more integration, but it did not yet result in a joint approach across all stakeholders. Strengthened coordination for the management of emergency situations through the CVC is now well established. Since 2009, risk management actors have met regularly in various steering committees and working groups on co-operative projects for integrated risk management. At the same time, the creation of the Fund to Combat the Effects of Natural Disasters (FLCN), under the direction of the Ministry of the Interior and financed by the government budget, has provided a specific financial tool for risk management. Its steering committee brings together the Secretaries General or representatives of the Ministries of the Interior, the Economy and Finance, General Affairs and Governance, and Urban Planning and Regional Development. The FLCN first covered emergency response and reconstruction, before extending to prevention issues. Its leadership in financing prevention projects also facilitates multi-player coordination of risk management (see Chapter 6).

A shared understanding of the remaining obstacles is gradually being established, including the need to better coordinate different activities. There is also growing recognition of the need for a truly multi-risk and multi-sectorial national risk management policy. The Ministry of General Affairs and Governance is responsible for developing a national risk management strategy and for ensuring its coordination with all national institutions and organisations that are involved.

Multi-level governance and empowering local players

These advancements have not yet been sufficiently transferred to the regional level, where municipalities that have been granted significant responsibilities, continue to face difficulties in implementation and weak accountability mechanisms. The involvement of local governments varies by region and the dynamics at work remain unclear. Provincial Flood Prevention Commissions, envisioned in the Ministry of Housing and Urban Development's Memorandum 824/2173 of 20 February 2003, bring together central and regional government officials to identify areas that are exposed to flood risk, to define next steps and to coordinate the implementation of these prevention measures. However, this interaction is uneven across the provinces and is often ignored, except in times of crises or reconstruction. *Ex ante* approaches are exceptional, but they seem to be productive, as evidenced in El Haouz province.

The predominance of top-down approaches in the regions does not empower local players to effectively implement and monitor risk prevention measures decided at the central level. This issue of empowerment, particularly in urban development and construction, is intimately linked to the development of a risk culture and to making decision makers accountable, as they deal with public opinion. It seems the approach of the current Community Action Plans (formerly Community Development Plans) that include local risks has not yet yielded meaningful results. Generally speaking, it seems the limited flexibility and resources made available to regional entities are hampering their empowerment at this level. However, at the regional level they remain important actors, particularly for prevention. The adequate connection between driving a strategic direction and its implementation on the ground is suffering from a lack of local involvement.

The decentralisation process currently under way in Morocco represents an opportunity to adjust vertical coordination between different government levels and to strengthen consultation mechanisms. Regions' and *provinces'* responsibilities in terms of land development, environmental protection and water management have been confirmed and reinforced by the decentralisation process. This level of government may be particularly appropriate for a local approach to integrated risk management if the process of decentralisation is accompanied with the mobilisation of the necessary human and financial resources.

Inclusion of civil society, businesses and the scientific community

In Morocco, the participation of civil society, businesses and the academic and scientific communities in the development and implementation of risk management policies varies, as there is no dedicated forum or platform to supervise their involvement. The absence of a dedicated space, where stakeholders can define a common vision, contribute to designing policies, share experiences, refine priorities and launch new initiatives, is hindering integrated action. However, several initiatives have emerged in past years and they deserve to be supported through a system of incentives.

Civil society

Morocco has a significant number of organisations (over 120,000), but they are not very involved in risk management issues. Civil society in the country lacks a specific structure for building partnerships within the risk management system. One exception is the Moroccan Red Crescent, which has been recognised as an entity of public interest and which has been involved in emergency response in the country since 1957. It supports government activities with rescue workers in explicitly selected aid posts and supports 300 aid centres.

In Morocco, civil organisations are governed by *Dahir* 1-58-378 of 15 November 1958, as supplemented and modified, regulating the right of association. This allows associations of persons to freely form civil organisations (Article 2) and establishes public charity status. Some associations play an important role in risk prevention, but their impact doesn't fulfil the potential of Morocco's associative movement. Among the successes identified by OECD peers, one can cite the Souss Life and Earth Sciences Association, which promotes seismic risk awareness in schools in the Agadir Region. Additionally, pilot projects for developing volunteer networks have been very successful in Fez and Meknes. Increasing the involvement of Morocco's associations will require creating incentives, as well as a formal structure for integration. The fact that Morocco has embarked on a larger national discussion with civil society should facilitate such an integration (OECD, 2015).

Businesses

In the absence of an inclusive risk governance structure in Morocco, private enterprises are not involved in the consultation processes on risk policies. The few themes that have emerged are those associated with business resilience and continuity of activity, even though businesses suffered significant economic damage due to the improper location of industrial districts, during the 2009 Casablanca floods, for example. However, some businesses now address the issue of risk in their activities. Continuity of activity in the banking sector is strongly supported by Bank-Al-Maghrib. Maroc Telecom applies international best practices to ensure service continuity. These best practices should be shared among the entire private sector community, and particularly with small and medium-sized businesses that are more vulnerable to critical risks.

In some cases, the government has established partnerships with common goals with the private sector. That is particularly the case for contractors that manage several key areas of critical infrastructure in Morocco. Law no. 54-05 on the contract management of public services, which was passed in 2006, includes the issue of service continuity and requires contractors to ensure services operate at the highest levels of security and environmental protection (Article 3). LYDEC, one of the largest water management contractors in Morocco, has committed to improve its accountability for risk. It conducts prevention operations and simulations.

Scientific community

Scientific and technical institutions, such as the National Meteorological Department, the National Centre for Scientific and Technical Research (which includes the National Geophysics Institute), and the Royal Institute for Remote Sensing (CRTS) participate within the risk management system through their scientific expertise and their technical resources. In particular, they support the identification and assessment of risks and early warning systems. The academic community's expertise is also used for specific projects. Universities regularly publish reports on risk-related topics, although they are not sufficiently integrated into the development of public policies. For example, they are not included in work on the risk of tsunamis, which will be addressed below.

Action plan to improve risk governance in Morocco

An opportunity

Efforts undertaken over the past 10 years have resulted in several initiatives to increase Morocco's resilience to the effects of critical risks. However, harmonising these efforts will require the development of a strategic framework that includes a holistic approach to the governance of critical risks and includes more risk prevention rather than crisis response.

These opportunities for the concretisation of such a policy in Morocco should be seized. A strong political will following significant flooding in November 2014, the current decentralisation process and its implications for regional governments, the development of important critical infrastructure projects, the public's expectation for strengthened public governance, and the 22nd session of the Conference of the Parties on Climate Change in 2016 can all contribute to this.

Governance gaps to be filled

Box 2.1. Use the Sendai Framework for Disaster Risk Reduction 2015-2030

The Sendai Framework for Disaster Risk Reduction 2015-2030 was adopted by 187 countries, including Morocco, during the United Nations 3rd World Conference on Disaster Risk Reduction, held 14 to 18 March 2015 in Sendai, Japan. Its goal was to substantially reduce loss of human life and property damage caused by natural disasters between now and 2030. The Sendai Framework includes forecasting the appearance of new risks and reducing existing risks by implementing economic, social, cultural, environmental, technological, political and institutional, health and education measures in an inclusive and integrated manner. These measures should reduce exposure to risks and vulnerability to disasters and improve the level of preparation for emergencies and reconstruction activities, thereby strengthening resilience.

The Sendai Framework emphasises a multi-risk, multi-player approach and calls for the integration of risk prevention in all public policies. Seven targets and four strategic priorities were defined to achieve this objective. The seven targets are:

- Significantly reduce the consequences of disasters in terms of mortality between now and 2030 (100,000 fewer deaths on average compared to 2005-2015).
- Significantly reduce the number of people affected between now and 2030 (100,000 fewer persons affected on average compared to 2005-2015).
- Reduce direct economic losses resulting from disasters between now and 2030.
- Significantly reduce the impact and damage caused by disasters to critical infrastructure, including damage to education and health, between now and 2030.
- Increase the number of countries that possess a risk management strategy between now and 2020.
- Improve international co-operation for developing countries between now and 2030.
- Strengthen early warning systems and improve risk information between now and 2030.

To achieve these objectives, the following four priorities were identified:

- Understand the risk of disasters.
- Strengthen governance mechanisms for disaster risk to ensure improved risk management.
- Invest in disaster risk reduction for resilience.
- Improve preparedness for an effective response, and encourage "build back better" in recovery, rehabilitation and reconstruction.

Source: UNISDR (2015), Sendai Framework for Disaster Risk Reduction, http://www.preventionweb.net/files/43291_sendaiframeworkfordrren.pdf.

Finally, the increase in initiatives and actors demonstrates a real energy around risk management in Morocco over recent years. Effective public policy tools have been developed to strengthen the country's resilience. Concerning governance, some gaps remain. Policies are too sector-based, single-risk, top-down and insufficiently inclusive and they prioritise emergency response and structural solutions. There is also a risk of inefficiency (duplication, transaction and opportunity costs) and a dispersion of resources. Current efforts cannot fully succeed. The imprecise allocation and inclusion of responsibilities and resources hinders the emergence of a common strategy (box 2.1).

The consultation process' gradual improvement and the consistency of legislation at the level of the central government may lead to the creation and implementation of a strategic vision. This vision could be constructed on two pillars: the CVC to coordinate emergency management, and the Fund to Combat the Effects of Natural Disasters (FLCN) as the principal financial tool to fund prevention policies. The establishment of a risk management strategy in Morocco could draw inspiration from several models identified by the OECD's High-Level Risk Forum. These models are presented later in this chapter.

Establish risk management objectives

First of all, a strategy must establish objectives for risk management. The OECD's Recommendation on the Governance of Critical Risk encourages countries to establish such objectives for all phases of the risk management cycle. Several approaches are used in OECD countries for this purpose. Some countries use a quantitative approach to risk, based on acceptable or tolerable levels of risks, depending on common preferences (OECD, 2104) and that have marginal costs, and aiming for an optimal level of risk reduction (Box 2.2). This allows protection standards to be set. In Switzerland, the definition of the acceptable risk level has set the probability of the loss of a human life at 10^{-5}. For economic losses, Switzerland uses the marginal cost concept, setting investment in prevention at the economically optimal level. The European Directive on floods sets three flood risk levels for which measures must be taken: low, medium, and high. High risk is the 500 to 1000-year flood level.

Box 2.2. Optimal investment in risk management

In risk management theory, optimal prevention measures are taken to maximise benefits for a given cost. Starting with the existing risk level, a utility curve can be defined to represent optimal prevention measures based on common preference. Since zero risk does not exist, prevention measures become more and more costly per benefit accrued. Benefits tend to diminish as the risk level is reduced. The marginal cost of prevention measures tends to increase up to the point where the cost-benefit relationship reverses. Investment in prevention becomes less profitable at this point, before it is no longer profitable at all. A traditional estimate is located at a point between one third and one half of the initial risk.

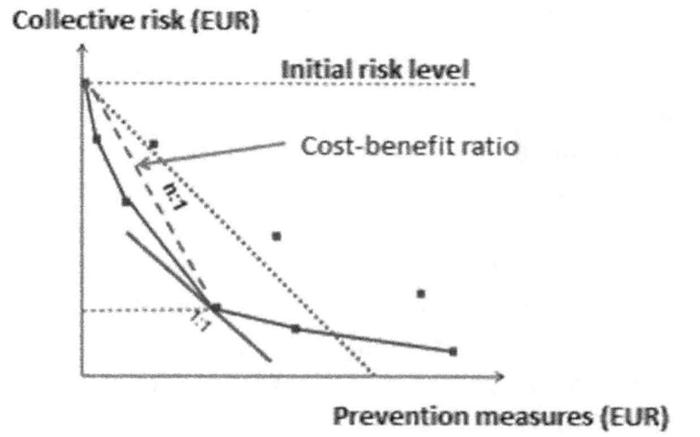

Source: OECD (2014), *OECD Study on the Management of Flood Risks: the Seine in Paris Region 2014*, OECD Publishing, Paris.

Objectives can also be set using a qualitative approach. In France, the flood risk management strategy adopted in 2015 set three objectives: (i) increase the population's safety, (ii) reduce the cost of damage and (iii) significantly shorten the time for affected areas to return to normal. In the United States, the five missions and 32 capacity areas for each actor in the national system are focused around a national preparedness objective. The goal is to become a "secure and resilient nation with the necessary capacities in the entire community to prevent, protect, mitigate, respond and rebuild from most risks and threats." These approaches could inspire Morocco in the development of its national risk management strategy.

Given the profile of risks covered by this study, it seems that, in the short term, Morocco could establish quantitative objectives, including: a protection standard for the risk of floods to provide guidance for prevention projects; objectives aimed at minimising the loss of human life related to the risks of earthquakes and tsunamis; an objective related to the economic impact of the risk of drought.

What should be the institutional framework for risk management?

Beyond setting objectives, the question of what would be the most appropriate institutional framework for developing and implementing such a strategy should be asked. Morocco now has two well-established pillars for risk management, allowing it to respond to crises strategically and in a coordinated manner: the CVC, and for decisions regarding the allocation of financial resources for prevention, the FLCN. Placed under the solid leadership of the Ministry of the Interior, which coordinates these two entities, this structure provides an excellent basis on which to build an integrated framework. Once it is adopted, the insurance system currently under development by the Ministry of Economy and Finance should complete such a structure. Two important elements are still missing in this institutional framework to allow it to cover the entire risk management cycle and respond to the principles articulated in the OECD Recommendation on the Governance of Critical Risks: a system for risk assessment (including "the capacity for prospective analysis, risk assessment and early detection that directly support rapid decision making") and a mechanism for the guidance and evaluation of all risk management policies ("leadership at the national level to conduct the implementation of public policies, connect the various sector polices and align priorities between the ministries and the different administration levels").

Several examples from OECD countries could be considered as models to fill the gaps in Morocco's risk management institutional framework. In response to significant duplication of responsibilities among various institutions responsible for risk management, Turkey reformed its system and created a single institution under the auspices of the Prime Minister (Box 2.3). Mexico chose a system based on three pillars: the coordination of emergency response, a financing mechanism for reconstruction and prevention, and a scientific centre of excellence for risk assessment and technical support. In France, there are two Ministries in charge: all aspects of prevention are under the responsibility of the Ministry of Ecology and Sustainable Development, while emergency management is under the authority of Ministry of the Interior.

Box 2.3. Institutional approaches to risk management in France, Mexico and Turkey

Several OECD countries have developed various institutional structures for risk management in order to improve coordination among actors and to ensure that all stages of the risk management cycle are covered. These approaches differ according to the institutional model of each country. Three examples may be sources of inspiration for Morocco:

In **France**, risk management is based on two strong pillars, the Ministry of the Interior for all policies related to emergency management and the Ministry of Ecology and Sustainable Development for prevention policies. The latter is also in charge of water policy, land development, public works and climate, and for ensuring good coordination on risk prevention. The government establishes the level of risk and requires municipalities to establish Risk Prevention Plans. Mayors are also responsible for the security of their residents and ask for support from the national government in times of crises, based on the subsidiarity principle. A compensation scheme that is established with the insurance sector provides the funding necessary for prevention measures. This system operates based on requests for proposals submitted by local actors.

In **Mexico**, the national civil defence system is supported by an office of general coordination under the Ministry of the Interior, which consists of three complementary entities: the civil defence department provides the link with state civil defence entities, the department of natural disaster funding provides financial resources for reconstruction and prevention by requesting proposals, and the National Centre for Disaster Prevention is the scientific and technical centre of excellence that develops tools and methodologies and gathers all the data and information related to risk. A National Emergency committee provides coordination in cases of major crises, and a National Civil Defence Council chaired by the President of Mexico meets annually to deliver strategic guidelines.

In **Turkey**, a major reorganisation of risk management took place in 2009. This reorganisation brought together three institutions into a single authority under the authority of the Prime Minister, known as the chair of disaster management (AFAD). Previously, the lack of coordination between the Ministry of the Interior's civil defence department, the Ministry of Public Works' disaster department and the Prime Minister's emergency management department led to significant deficiencies that were revealed during an earthquake in the Sea of Marmara in 1999. This transformation allowed the country to move from a disaster management approach to an integrated risk management approach. AFAD is now in charge of defining risk management policies, establishing standards, financing projects, providing emergency response, and coordinating all risk management activities.

Sources: OECD (2013), *OECD Reviews of Risk Management Policies: Mexico 2013: Review of the Mexican National Civil Protection System*, OECD Publishing, Paris; OECD (2014), *OECD Study on the Management of Flood Risks: the Seine in Paris Region 2014*, OECD Publishing, Paris.

Ensure coordination: defining roles and responsibilities

A key issue when it comes to coordinating different levels of government and strengthening the inclusion of regional authorities, civil society, businesses and the scientific community in the definition and implementation of public policies is the lack of a sufficiently inclusive structure. Various stakeholders have difficulties agreeing on the overall objectives of risk management. Consequently, there is neither coordination between, nor alignment among the various initiatives. OECD countries have used several approaches that could inspire Morocco's efforts. A good method for clarifying roles has been developed in Finland. Starting with identifying and analysing all critical functions that must work under all circumstances, Finland assessed how these would be impacted by different risk scenarios, before establishing a responsibility map of each function for each institution (Box 2.4). In Mexico, a manual was developed that precisely specifies the role of each entity, allowing all stakeholders to mobilise around the question of risk management (Box 2.5).

Box 2.4. Finland's national security strategy, a methodology for strategy building

The Finnish Strategy for the Security of Society is built on a pan-governmental approach to security with the goal of securing all of society's vital functions in all possible situations. The strategy aims to make Finland the most secure country in Europe. Promulgated in 2010 as a Government Resolution, this is a strategic, multi-risk initiative that provides a uniform basis to ensure the preparedness of all Finnish institutions, from the central level to municipalities, and the private sector and NGOs. The strategy was prepared by the Secretariat of the Security and Defence Commission and by managers of each ministry. The strategy determines which vital functions of Finnish society must be protected in all situations. The international dimension of dangers and threats, specific vulnerabilities associated with disruptions of critical infrastructure and dependence on the continuity of global supply chains are among the key elements included in this strategy. In addition, it enumerates the strategic tasks that contribute to ensuring vital functions and it assigns each of these responsibilities to a ministry. Possible disruptions in the completion of these strategic tasks and in the functioning of vitally important sectors are illustrated in different scenarios that provide a foundation of preparedness levels for all stakeholders. Finally, the strategy defines preparedness and principles of crisis management.

Source: OECD (2016), Toolkit on Risk Governance, https://www.oecd.org/governance/toolkit-on-risk-governance/, reviewed 12 February 2015.

Box 2.5. Mexican national civil defence system manual

Mexico's national civil defence system has gradually developed since the 1985 earthquake, and today features an integrated risk management model. Initially established to improve the organisation of the emergency response system, which demonstrated serious deficiencies, it has progressively developed a legal and institutional structure for the entire risk management cycle. The publication in 2006 of an operational manual represented a significant step in this implementation procedure.

This new approach is based on the need to precisely assign roles and responsibilities to each stakeholder. It resulted in the publication of the SINAPROC organisational and operational manual in 2006. This manual defines for the first time the mandates of all players in three key areas of risk management: prevention, intervention and recovery. Although the organisation and entities within SINAPROC remained unchanged, the functions and responsibilities of the actors involved were, for the first time, presented in a detailed and precise fashion. SINAPROC stakeholders included federal ministries and entities, the private sector and civil organisations, such as the Red Cross, but also the media and professional organisations in the chemical industry.

The manual played an important role in involving all federal institutions and making risk management a priority for all relevant entities. With 38 organisations active in prevention, 34 in the emergency phase, and 18 in recovery and reconstruction, coordination is essential. The manual defines the various roles for each of the following areas: executive coordination, technical coordination, technical support and shared responsibility.

Box 2.5. Mexican national civil defence system manual *(continued)*

SEGOB remains the executive coordinator, while the states and municipalities are the technical coordinators, with some exceptions. However, although the manual defines the roles and responsibilities of each actor, it does not specifically address coordination beyond the existing Civil Defence Office and National Emergency Committee.

Source: OECD (2013), *OECD Reviews of Risk Management Policies: Mexico 2013: Review of the Mexican National Civil Protection System*, OECD Publishing, Paris.

Responsibility and accountability

Beyond setting specific objectives, improved player responsibility can be established through developing accountability mechanisms. This includes regular reporting on the progress towards the strategy's implementation, and establishing a follow-up mechanism and a forum for public accountability.

In Morocco, despite the involvement of a large number of actors, there are no criteria for evaluating how measures taken by various actors contribute to managing risks. This lack of performance evaluation makes it difficult to assign risk management responsibilities and resources in an optimal way. Current mechanisms to accompany risk management policies by assigning responsibility, monitoring, providing incentives and ensuring stakeholder accountability still appear insufficient in Morocco.

The European Directive on Floods is a good example of a cyclic method, where risk management objectives are regularly reviewed (Box 2.6). In Norway, an institutional structure dedicated to assessing risk management policies has been established and the Ministry of Justice conducts an annual audit of this policy's implementation in all ministries (Box 2.7). In Turkey, the Court of Auditors has an audit system dedicated to disaster management. Turkey has also actively contributed to the development of an International Organisation of Supreme Audit Institutions standard for risk management audits (ISSAI 5510).

Box 2.6. The European Directive on Floods and its cyclic approach

European Directive 2007/60/EC on Floods dated from 23 October 2007 aims to reduce the negative consequences of floods. It considered changes in risk factors and ensured the follow-up and evaluation of strategies implemented to manage the risk. It requires member states to identify and map at-risk areas and establish flood risk management plans. This Directive's goal is to concentrate prevention efforts in the most exposed areas by putting in place rigorous and quantifiable objectives to reduce vulnerability, focused on the potential impact of floods on human health, the economy, the environment and cultural heritage. While providing member states with some flexibility, the Directive establishes a three-step strategy with a clear schedule. The strategy includes:

- A Preliminary Assessment of the Risk of Flood (EPRI) for major river basins and nationally to identify areas that are most at risk;

Box 2.6 The European Directive on Floods and its cyclic approach *(continued)*

- The development of maps identifying flood zones and flood risks for low, medium and high probability floods in risk areas;
- The development of a Flood Risk Management Plan (PGRI) for each river basin with management objectives and specific measures to achieve them.

These steps are to be reviewed every six years.

Source: OECD (2014), *OECD Study on the Management of Flood Risks: the Seine in Paris Region 2014,* OECD Publishing, Paris.

Box 2.7 Risk management policy assessment in Norway

In Norway, the Ministry of Justice and Public Security conducts audits to guarantee a national standard for preparedness within each ministry. Each ministry that is responsible for the day-to-day management of a particular sector (such as energy, communications or transport) is also accountable for prevention, planning emergency measures and implementing an emergency crisis response. The Ministry of Justice and Public Security plays a role in coordinating comprehensive and coordinated civil defence activities. The Ministry of Justice and Public Security monitors the activity of different ministries with civil defence, as well as their level of preparedness. These audits aim to ensure preparedness for each specialised ministry by guaranteeing the quality of emergency response planning, by facilitating a targeted and efficient use of resources, and by contributing to the assessment of cross-sector needs. The first audit was conducted in 2002. Regular audits began in 2005, and they are now completed regularly.

Source: OECD (2016), Toolkit for Risk Governance, https://www.oecd.org/governance/toolkit-on-risk-governance/, reviewed 15 February 2016.

Conclusion and recommendations

In the past 20 years, Morocco has made significant progress in critical risk management. This progress is visible in terms of governance improvements through a body of legislation and regulation, programmes and master plans, the establishment of dedicated financial mechanisms, the assignment of responsibility to diverse entities *sui generis,* as well as sector-based economic development plans that integrate the issue of risk up to a certain point. This progress has been accompanied by several pilot projects led by civil society organisations, decentralised government authorities, and public-private partnerships. At the time of the OECD peer mission to Morocco, almost 45 best practices have been documented (Appendix C).

This effort towards building a risk governance structure has benefitted from Morocco's growing integration into international value chains, the external financial and technical support it has been receiving, and the need to respond to a series of significant catastrophic events. The "response" dimension, as evidenced by the importance of crisis and reconstruction management, represented a driving force behind the creation of this structure.

While these efforts are laudable, in the absence of an integrated governance response structure, they have also allowed the development of fragmented risk management elements. Clear governance shortfalls remain, as policies are too sector-based, single-risk, top-down, insufficiently inclusive, and favour emergency response and structural approaches. There is a risk of inefficiency, dispersion of resources, and that efforts fail to reach their full potential. Poorly assigned and poorly understood responsibilities and resources hinder the emergence of a common strategy.

The current management system's vulnerabilities should be assessed using a "gap analysis" that determines the unmet challenges for a fully functional risk governance structure over the short, medium and long terms. Is the time horizon of current forecasting and planning activities sufficient? Does governance sufficiently cut across various policies? Can the coordination of various actors be optimised? Does the system sufficiently integrate vertical levels of government, including regions, provinces and municipalities? Can civil society actors truly be considered stakeholders? How can resource allocation be further improved between sectors, within the risk management cycle, and for different time horizons?

The key issues relate to the coordination between different levels of government and the wider inclusion of regional entities, civil society, businesses and the scientific community in policy setting and implementations, which are lacking in the absence of a sufficiently inclusive structure. Various stakeholders have difficulty agreeing on overall risk management objectives. As a consequence, there is no coordination nor alignment of their respective initiatives with a common strategy. Despite the involvement of a large number of actors, there are no criteria to assess the contributions of their measures. This lack of performance evaluation of different stakeholders' initiatives heightens the challenge of optimally assigning responsibilities and resources for risk management. However, consultation processes on this subject and central government legislations have progressively improved and become more consistent. It is necessary that Morocco speeds up the process in order to establish a strategic vision.

As Morocco has reached a greater understanding of risk management, it is now an opportune time to capitalise on the current situation and prepare for tomorrow's governance. Since several questions are long-term oriented (urban development, water, transport and energy infrastructure), and based on social factors such as demographics, educational level, risk and culture, the time frame for risk to be taken into account in the various traditional sector plans must be between 50 and 100 years. Similarly, considering the impacts of climate change or the occurrence of a tsunami in Morocco can only occur within very long time frames.

Aligning governance structures with related challenges and longterm horizons requires a heightened level of expertise. Moroccan stakeholders and OECD expert groups and peers firmly believe Morocco has now acquired the capacity to increase its ambition level and that it can build such a governance structure serving an integrated risk management strategy, based on a long term vision.

Recommendations

BUILD A LONG TERM VISION FOR AN INTEGRATED GOVERNANCE STRATEGY FOR CRITICAL RISKS

Set the fundamental principles of a long-term vision for risk management in coherence with Morocco's 2030 target, and rank objectives accordingly.

Based on an assessment of shared risks, the strategy will establish and rank the principal objectives with follow-up and assessment indicators based on a timetable for action. The definition of a flood protection level (e.g., hundred year flood), a goal aimed at achieving "zero deaths" following an earthquake or a tsunami, and raising awareness of the economic impact of droughts could represent viable objectives for such a strategy.

Grant national risk policy a status of national priority by assigning a dedicated space in the institutional organisation chart that is representative, and provide it with a certain leadership.

To provide sufficient credibility an integrated critical risk management strategy must be visible within the country's institutional organisation chart at a very high level, not just as a symbol. In some countries, this has been achieved by creating an interministerial department, a High Commission, a dedicated ministry, or a chaired interministerial committee.

Establish an adapted institutional structure for the governance of a national priority with clear lines of responsibility, covering the entire risk management cycle.

This involves building on the existing structure for risk management in Morocco; the CVC for crisis management and the FLCN for the strategic allocation of prevention resources. Establishing an additional independent structure responsible for continual risk assessment would fill a significant gap. A structure dedicated to the assessment of the national strategy's implementation must also be developed in order to guarantee adequate accountability.

Modernise the regulatory framework and adapt it to current and future challenges.

The current revision of the legislative corpus and new laws must take risk management into account. In particular, this relates to the law on compensation mechanisms, the law on shorelines, the application laws of the sustainable development charter, as well as revisions of the water law, critical infrastructure legislation and urban development requirements. Specifically, significant efforts must be channelled towards creating a legal framework for urban development guidelines in flood, seismic or tsunami zones.

Coordinate the implementation of risk management policies, align them with national sector development strategies, and create bridges with regional and local actors where appropriate.

All development plans and sector strategies in Morocco must integrate a risk dimension related to the national critical risk management strategy. This must include all regional development plans, both in regions and municipalities, and the country's various socioeconomic strategies. The law on sustainable development's implementation would represent an opportunity to deliver on this recommendation.

Create and lead a forum that discusses the major action points of risk management, allowing citizens and civil society to get involved.

A neutral space could allow a wide-ranging exchange to be held between national entities, elected officials, communities, private sector actors and civil society. Such a forum would also serve to report on progress achieved towards implementing the country's critical risk management strategy. Regular discussions or conferences on risk could also be explored.

Bibliography

World Bank (2011), "Phase 1: Assessment of Risks in the Current Situation and to the 2030 Horizon for the City of Casablanca and the Bouregreg Valley" in *Adaptation to Climate Change and Natural Disasters by Coastal Cities in North Africa*, World Bank.

World Bank (2015), "Assessment of Environmental and Social Systems in Morocco" in *Integrated Risk Management Programme for Natural Disasters and Resilience,* World Bank.

Kurt, S. (2014), From Theory to Practice on Governance, presentation at the OECD workshop on "Learning from Crises and Fostering the Continuous Improvement of Risk Governance and Management", Oslo, 17 September 2014.

Ministry of Water, Assessment of the Implementation of the National Flood Protection Plan, Ministry of Water.

Ministry of the Environment (2010), National Water Strategy, http://www.environnement.gov.ma/PDFs/EAU/STRATEGIE_EAU.pdf.

Ministry of General Affairs and Governance (2014), Towards an Integrated Risk Management Strategy in Morocco, presentation during interviews for the OECD Study on Risk Management in Morocco, Rabat, 20 February 2014.

Ministry of Public Works, Transport and Logistics (2015), Prevention and management of flood risks within the METL, presentation during interviews for the OECD Study on Risk Management in Morocco, Rabat, 6 June 2015.

OECD (2013), OECD Reviews of Risk Management Policies: *Mexico 2013: Review of the Mexican National Civil Protection System*, OECD Publishing, Paris.

OCDE (2015), *OECD Examination of Public Governance: Open Government in Morocco.*

OECD (2014), *Boosting Resilience through Innovative Risk Governance*, OECD Publishing, Paris.

OECD (2014), *OECD Study on the Management of Flood Risks: the Seine in Paris Region 2014*, OECD Publishing, Paris.

OECD (2016), *Toolkit for Risk Governance*, https://www.oecd.org/governance/toolkit-on-risk-governance/.

PwC et al. (2014), National Sustainable Development Strategy, Final Report.

Chapter 3.

Risk assessment in Morocco

A precise and shared knowledge of the risks at stake constitutes the foundation for any decision relating to risk prevention and raising awareness about the development of risk culture. This chapter outlines the institutional context of risk assessment in Morocco, the recent changes it has witnessed. It addresses questions regarding measures, sources and access to strategic data that provide crucial information about various types of hazards and challenges facing the country. Finally, the chapter discusses how Morocco can take advantage of its skills base and pilot tools in order to establish an adequate risk assessment process that is both fully integrated with its strategy and operational.

Introduction

A precise and shared knowledge of the risks constitutes the foundation on which any decision relating to risk prevention and awareness raising is based. The assessment and knowledge of risks helps to methodically determine their nature and characteristics, particularly their magnitude, the probability of their occurrence, and their possible consequences.

The OECD Recommendation on the Governance of Critical Risks promotes a risk assessment approach based on a comparison of different risks, which helps to set priorities and inform various areas of public policy, from creating an integrated strategy, to prevention, response, and compensation measures. A systematic approach based on the best scientific knowledge available is necessary to ensure risk management policies and related investments are appropriately channelled (figure 3.1). A poor risk assessment can result in uninformed or arbitrary decision-making, in turn leading to costly and overly protective policies or to dangerous oversights that jeopardise the safety of people and property.

Figure 3.1. Risk assessment

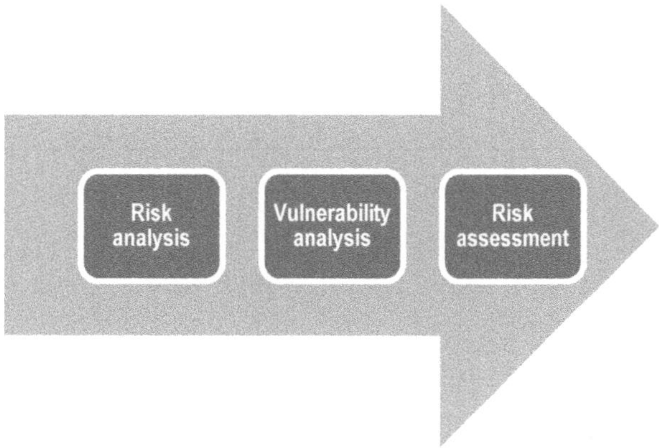

Background to risk assessment in Morocco

A number of initiatives have been launched in Morocco to assess critical risks since the beginning of the 2000s. However, there is no precisely established framework assigning related roles and responsibilities of risk assessment. As a consequence, numerous institutions take part in various risk assessment processes, based on their respective area of intervention, the information they possess about the risk at hand, and their scientific or technical expertise.

Following conventional lines, Morocco adopted single-risk approaches to assess major national risks at the start of the 2000s (flooding and earthquakes). Since the 2004 earthquake, it has also strengthened multi-risk assessments aimed at creating better emergency responses. The last few years witnessed the development of multi-risk approaches that offer a more integrated view of the entire risk management cycle (Table 3.1).

Table 3.1. Main risk assessment processes in Morocco

Type of assessment	Institution responsible	Date	Objectives
Single-hazard assessment			
Earthquake risk assessment	Ministry of Housing and Urban Development	2004 (revised in 2008)	Regulation →Development of paraseismic regulations
Flood risk assessment	Ministry of Water	2002 (under revision)	Prioritisation of investment →Development of National Flood Protection Plan
Multi-hazard assessments			
Risk monograph	Monitoring and Coordination Centre/wilayas	Since 2004	Emergency action →Development of local emergency plans
Master Plan for Risk Analysis and Coverage	Directorate-General of Civil Defence	Since 2004	Prioritisation of investment →Planning of emergency response resources
National risk atlas	Ministry for the Environment	2008	Communicating the risk →Mapping and raising awareness
MnhPRA	Ministry responsible for governance	2012	Investment →Prioritisation of preventive action Regulation →Risk financing strategy
Database			
Participation in DESINVENTAR (UNISDR) database	Ministry for the Environment	Since 2010	Gathering of information on loss of human life and property

Source: interviews conducted by the OECD

Single-risk assessments

Risks were traditionally assessed individually by the various institutions responsible. Accordingly, the 1995 Law on water allowed the development of a National Flood Protection Programme based on an assessment of flood risk. The Ministry of Water then carried out a flood risk assessment in 2003. This enabled a nationwide identification of different levels of risk by combining the rate of water flow with a typology of vulnerable areas, based on a risk matrix. Subsequently, 391 high-risk areas were identified, among which 50 areas were assigned a high-priority status. However, according to the experts and stakeholders, this national approach is still not sufficiently detailed and it lacks local-level assessments from Hydraulic Basin Agencies. Revisions to this plan are ongoing.

Likewise, seismic risk mapping has led to the establishment of national regulatory zoning, a courtesy of the Rabat Scientific Institute database. In fact, the Institute is the first body to gather seismic data since the beginning of the 20th century. However, the information collected is not detailed enough to be used for estimations of potential damage. More accurate micro-zoning studies remain limited. There is currently no assessment approach for building and infrastructure resistance, and no relevant database to inform such an approach.

Risk assessment for emergency planning

Various bodies of the Ministry of the Interior are also responsible for developing risk analyses regarding emergency response planning. The Monitoring and Coordination Centre, which was set up in 2004 after the Al Hoceima earthquake, is responsible for developing national emergency plans and their corresponding risk analysis. *Wilayas* must organise monographs of risks affecting their region in order to plan the risk management strategies to be implemented in each region. Meanwhile, the Directorate-General of Civil Defence must prepare the planning of emergency resources by drafting Master Plans for Risk Analysis and Coverage (see Chapter 5). These different initiatives are often enshrined in the legislation. However, the methodology to be used is not always clarified in detail, which makes it possible to put in place various types of risk assessment. Therefore, master plans for risk analysis and coverage such as monographs usually fail to gather all the information available on historical risks from different administrations, and they do not apply precise and focused methodologies.

Moving towards integrated multi-risk approaches

The first multi-risk atlas appears in 2008

In line with the growing momentum surrounding risk management during the 2000s, the Ministry for the Environment published the first atlas of critical risks in Morocco in 2008. The atlas compiled a collection of maps related to the risks of earthquakes, floods, tsunamis, forest fires and landslides, as well as risks of industrial origin. This led to the development of a Geographic Information System gathering all of the collected information on behalf of various scientific and technical institutions, based on their abilities and responsibilities. The first integrated national risk assessment database was subsequently established. However, the atlas only included information on principal hazards and did not comprehensively address questions relating to the vulnerability to these risks or about their potential damage. Summary reports covering the principal risks were drafted, but these different elements have not been updated since due to difficulties with the supply of data provided by key partners.

Multi-risk assessment is probabilistic and ambitious but under-used

From 2008 onwards, as part of the efforts towards the development of an integrated risk management strategy, the Ministry of General Affairs and Governance became involved in developing a high-performance risk assessment tool to include both issues of exposure and vulnerability. To this end, a partnership with the World Bank was forged to develop an ambitious technical tool that would combine a probabilistic analysis of principal risks with their potential impacts (Box 3.1). Completed in 2012, the tool has rarely been used or applied due to a lack of training on how to actually use it, according to various Moroccan stakeholders. In addition, a low level of appropriation of this high-performance tool seems to have prevented it from being optimally used. Some criticisms have pointed to the "black box" aspect of the tool, which demonstrates the critical need to fully mobilise local technical and scientific competencies to face such challenges.

Box 3.1. MNhPRA, an effective but under-used risk assessment tool

Efforts to analyse different hazards facing Morocco were recently strengthened through a more integrated project run jointly by the Ministry of General Affairs and Governance and the World Bank, which culminated in the development of the MnhPRA (*Morocco Natural Hazards Probabilistic Risk Analysis*), a GIS-based software that runs models assessing the economic impact of four principal risks, namely floods, earthquakes, tsunamis and droughts. A nationwide list of properties exposed to risk (including residential housing and infrastructure) helped carry out a probabilistic study to estimate incurred losses incurred due to each principal risk, ranking them according to the frequency of occurrence, across Morocco's entire land mass, and in selected regions, provinces and municipalities. A calculation model showing the macroeconomic impact of the different risks was also developed within this framework, along with a model assessing the costs and benefits of various preventive measures.

While this approach helped overcome challenges linked to cooperation and exchange of data, it also revealed certain limitations: a number of stakeholders, particularly those in scientific and technical fields, did not feel sufficiently involved, and their appropriation and use of the tool remain low to this day. Moreover, the MnhPRA doesn't account for hazards' impacts on human populations, nor for their indirect effects on critical infrastructures. Nevertheless, it represents a highly useful database for the Ministry of Economy and Finance, in particular for the development of a damage compensation system.

Source: World Bank (2014), *Strengthening of resilience in Morocco: input on an integrated risk management strategy*, Washington, D.C.

Local risk assessment

At the regional level, risk assessments are carried out based on *ad hoc* studies, which are not systematic given the absence of legal or regulatory requirements and the lack of incentives. There is a strong need for methodological support, provision and availability of information, and for strengthening the existing expertise. At the level of communities, risk assessment approaches remain limited. As a result, it is not guaranteed that such issues will be included in local development efforts.

According to various experts and stakeholders, numerous risk assessment approaches are adopted independently at the level of local authorities:

- Risk monographs carried out in the *wilayas* and provinces – laying out the groundwork for an emergency response – gather existing information on risks, without following a standardised methodology.
- Flood-risk atlases exist for some hydraulic basins, like the one in Haouz province, but they do not cover all of them. The detailed map, featuring 391 risk hotspots, is not exhaustive.
- Urban agencies also include elements relating to existing critical risks in their planning documents, principally in the form of a hazard map. These rarely identify the level of risk according to the associated challenges and vulnerability, although it would facilitate setting more comprehensive public planning policies. Such an integration is not carried out systematically in the absence of specific regulations on this matter.

Ad hoc studies are also carried out as part of international cooperation activities such as development and infrastructure projects and assistance for local development. These assessments sometimes adhere to high international standards (incorporating the concepts of vulnerability, economic assessment, and climate predictions), and they would merit being used more extensively. In this context, a 2012 pilot study conducted by the Ministry for the Environment on adapting to climate change and natural risks in Greater Casablanca and the Bouregreg Valley produced a detailed assessment of the risks of flooding and tsunamis for Casablanca's urban area. The activities carried out for the development of a high-speed train linking Tangiers and Rabat also respected the highest international standards in terms of assessment of earthquakes and flooding risks. The city of Tétouan also benefitted from international cooperation in order to develop a risk assessment process and a risk observatory (Box 3.2). Meanwhile, the city of Fez has been working closely with the Swiss Cooperation to develop highly detailed flood-risk hazard maps (Box 3.3).

Box 3.2. Local guide to natural risk prevention in Tétouan

In order to develop a risk culture and to raise awareness among communities in the Tangier-Tétouan Region, the Urban Municipality of Tétouan negotiated a four-party agreement with the Mediterranean Centre for Environment and Development (CMED), the Mediterranean Centre for the Environment (CME), and Provence-Alpes-Côte d'Azur Region (PACA). The agreement led to the creation of a Centre for Natural Risk Culture and Awareness at Faculty of Science in Tétouan. This newly established centre developed Morocco's first "Local guide to natural risk prevention."

The document lists and describes principal natural risks that Tétouan is exposed to (flooding, landslides, earthquakes and forest fires), and highlights various existing vulnerability factors. It also offers specific safety advice for each of the aforementioned risks. Brochures were produced and distributed in schools and to civil society representatives to help raise awareness among the general public.

This project, financed by the PACA Region, convened a large network of participants, including teachers, researchers and students at the Tétouan Faculty of Science and at Abdelmalek Esaâdi University in Tangier, as well as engineers from the Urban Municipality of Tétouan and the CME. A large amount of positive feedback was received, justifying the Municipality of Tétouan's approach to deal with natural hazards. It is expected to expand the mission of the Centre for Natural Risk Culture and Awareness, and to turn it into a "natural risk observatory".

Source: Urban Municipality of Tétouan (2013), Local guide to natural risk prevention in Tétouan, interviews conducted as part of the OECD Study on Risk Management in Morocco (May 2015).

Box 3.3. Mapping of flood risks in Fez and Beni Mellal

As part of international cooperation initiatives between Morocco and Switzerland, the Swiss Agency for Development and Cooperation (DDC) was commissioned to study the hydrological dangers in the river basins of Fez and Beni Mellal Regions.

Implemented in 2009-2010 and drawing on a total budget of CHF 250,000, this project aimed to publish a flood-risk map of the two river basins. It offered the opportunity to gather recommendations for stakeholders responsible for flood-risk management in Morocco, making the most of Switzerland's recognised experience in the matter. For that purpose, the Institute of Geography at the University of Lausanne (Switzerland) and Sidi Mohamed Ben Abdellah University in Fez (Morocco) created a practical guide to support the application of the Swiss method for mapping the risk of flooding in Morocco. The guideline is useful for practitioners (water managers, engineering consultancies and town planners) and it focuses on two selected case studies; the river basin of the Fez *wad* and the river basins from the Beni Mellal urban area.

By looking at regions with different hydro-geomorphological characteristics, this guide enables readers to think about various environments, which facilitates the implementation of recommendations and methodologies. In the case of Beni Mellal, the guide uses a "naturalist" approach, while an empirical approach is preferred for Fez. It also distinguishes between the 1D modelling of the "valleys" area in Fez, and the 2D modelling of the "piedmont cones" in Beni Mellal. Similarly, the flood-risk map is also based on Swiss practices, and combines analysis of a flooding phenomenon's maximum intensity, and the probability of its occurrence. Furthermore, intensity thresholds were established for each hazard under consideration.

Source: Werren, G. and M. Lasri (2014), *Practical Guide to Applying the Swiss Method in Morocco*, Flood-risk map, Institute of Geography and Sustainability at the University of Lausanne (IGD), Geo-environmental and Planning Test Laboratory at Sidi Mohamed Ben Abdellah University in Fez (LAGEA), https://serval.unil.ch/resource/serval:BIB_D261C5C97A38.P001/REF

Risk assessment: data availability and access

Establishing institutional mandates, including specific roles and responsibilities, is essential for the development of an effective risk assessment process. However, this requires the collection of large amounts of data on hazards,exposure to risk, as well as their integration into a risk analysis. Achieving this also requires an adequate concentration and dissemination of hazard measurement networks, and the ability to collect, process and compile multiple datasets, as well as technical capabilities to model the risk.

Meteorological hazard measurements and data

The National Meteorological Office in Morocco (DMN) considerably strengthened its measurement resources over recent years. The establishment of the Fund to Combat the Effects of Natural Disasters in 2009 helped raise 135 million Dirhams during the 2010-2012 period to modernise the meteorological service, as part of the VIGIOBS project. The number of automatic weather stations increased from 50 to over 200, while the number of radars rose from 6 to 8. In addition, 370 weather stations were run by other Moroccan institutions, such as the hydraulic basin agencies, the Ministry of the Interior, the High Commission for Waterways and Forests and the Fight against Desertification, and the National Water and Electricity Authority, which operates some of the country's dams. The DMN also has access to other measurement networks for high altitudes and at sea, radio transmission systems, and data processing capabilities.

Morocco can, therefore, count on a good coverage of its land mass, as well as necessary competencies to create a climate database that records temperature and rainfall data for the entire country. However, the high spatial and temporal variability of precipitation would require improvements to the measurement network to ensure a better knowledge of the risks of flooding, particularly in the mountainous basins with reactive flows. Morocco similarly possesses long-term datasets that date back to 1960, which have been archived, and the quality of which is monitored. This represents a solid basis to build on efforts to run models of the risks of flooding, drought, forest fires and landslides. The DMN has also invested in its modelling capabilities with regards to the effects of climate change, and it can consequently provide meteorological datasets for various climate change scenarios (Box 3.4) (Box 3.5). Although the data gathered by the DMN is shared with the main actors and operators of critical infrastructure, it is important to stress that this data is not yet freely accessible and it requires an agreement before it can be shared.

Box 3.4. National Meteorological Office's climate modelling capabilities

The DMN has invested in the development of its modelling capabilities vis-à-vis the effects of climate change in Morocco. It can use the results of global circulation models that simulate the climate on a planetary scale and scale down to Morocco's level. This makes it possible to obtain results that include the special features of the Moroccan climate, particularly the country's regional peculiarities; it is especially useful for modelling the effects of climate change on extreme phenomena such as drought and flooding.

Source: Interviews conducted as part of the OECD Study on Risk Management in Morocco (May 2015).

Box 3.5. Composite drought indicator

Regarding the risk of drought, the LDAS project recently launched by the Royal Centre for Remote Sensing (CRTS), in collaboration with the Hassan II Agronomic and Veterinary Institute (IAV), developed a Composite Drought Indicator (CDI). Its objective is to quantify the phenomenon of drought in a realistic manner by determining and agreeing on a set of criteria or on composite indicators. As part of this study, the criteria involved the Standardised Precipitation Index (SPI), soil humidity (LST), and evapotranspiration (ET), as well as the anomalies of the Normalised Difference Vegetation Index (NDVI). These four criteria or indicators are brought together in the CDI, which represents the weighted sum of the products of each indicator:

CDI= 0.4*SPI+ 0.2*NDVI+ 0.2*Lst+ 0.2*Et

Source: National Drought Mitigation Center (2015), Researchers collaborate with Morocco on state-of-the-art satellite-based drought index, NDMC news, 20 March 2015, http://drought.unl.edu/NewsOutreach/NDMCNews.aspx?id=193, consulted on 4 March 2016.

Water hazard measurements and data

The Ministry of Energy, Mining, Water and Environment's Department of Water possesses 265 hydrographic stations and 700 periodic measurement points. As such, the central administration and the nine Hydraulic Basin Agencies (ABH) collect data on flow and wave height on a regular basis. All of Morocco's dams provide an accurate evaluation of the flow of water released downstream into rivers, while water usage for irrigation purposes is also monitored and measured. This is an important source of hydrological and hydraulic information. It should, however, be noted that given the hydrology system of the Mediterranean's irregular water regimes, several *oueds* which are at risk of important flooding in the event of storms, are not equipped with measurement systems.

The Ministry and the ABHs publish an annual hydrological report that details the flooding and flow peaks of various rivers. These are available for certain years only, and can be accessed on the websites of the Ministry and most ABHs. Generally, these institutions have Geographic Information Systems (GIS), which are able to map all relevant aspects related to water resources. However, it does not appear that Morocco's river flow series are all gathered in a common database nor easily accessible for scientific and technical work relating to risk assessment. Rather, one can note a high diversity in terms of available data, its format, and the time periods it covers. The lack of consistency of available data is hindering the wider development of flood-risk assessment at all levels of Moroccan society.

Characterising the risk of flooding also involves the development of hydrological modelling for various river basins. Such models provide information on the return periods, frequency measurements, and the main characteristics of floods in terms of flow at different points of the basin and flood expansion, on the basis of observed rainfall or flow series over different time periods. Based on this information, a complete assessment of damage and losses for different flooding scenarios can be prepared. The development of flood zone modelling in Morocco is still in its early stages, despite the considerable efforts expended as part of the National Flood Protection Plan, which had identified the country's principal high-risk sites through the use of global modelling. A further detailed account of the risk is necessary to obtain more accurate information on the potential damage that can be caused by flooding. This requires geographical data such as digital models of mountainous areas, and hydrological models.

Earthquake and tsunami hazard measurements and data

Seismic risk monitoring is carried out by the National Geophysical Institute (ING) within the National Centre of Scientific and Technical Research. The Institute handles the design, installation, maintenance and operation of a network across the country. The network is composed of around 50 short-period analogue seismic stations, 50 new-generation digital stations, two thirds of which are broadband-enabled, and 50 strong-motion seismographslocated at the country's major structures. There is also a network of GPS stations in the planning stage, which will help measure shifts along earthquake fault lines and the accumulation of potential energy. As part of a partnership with the Ministry of Public Works' Ports Directorate, this has been complemented by the instalment of a network of tide gauges to measure and warn of the risk of tsunamis (Box 3.6). Another objective is to develop the network of strong-motion seismographs in large towns that are most threatened by earthquakes, thereby providing more accurate information about ground movement at local level.

The ING has developed an extensive database of earthquakes that struck the Moroccan land mass since 1900, featuring geographic location of epicentres and their corresponding magnitudes. Work led by national universities and international researchers has also helped create a database and analyse the main earthquakes that hit Morocco from the 9th century to 2004 (El Mrabet, 2005). There is thus a catalogue of historical earthquakes in Morocco along with a catalogue of major tsunamis that reached the Moroccan coast (F. Kaabouben et al, 2009). Based on a map of seismic hazards covering the entire country, a risk map was created to show peak ground acceleration and ground movement speed across five risk categories. In 2002, paraseismic regulations were established in Morocco. Supported by these maps, parasesimic regulations were revised in 2008, under the auspices of the National Paraseismic Engineering Committee.

For higher-risk areas, seismic micro-zoning studies can provide an accurate assessment of the risk of earthquakes based on a more detailed ground analysis, such as the one conducted in partnership with the French Cooperation in Al Hoceima Region after the 2004 earthquake. A similar project was later approved in 2005 to finance prevention initiatives in Agadir.

Box 3.6. Tsunami risk modelling

Often missing from prevention policies in Morocco, tsunamis remain nevertheless a serious threat to Morocco's major coastal cities, where the majority of the country's population and economic activity is concentrated.

As part the activities conducted by the Royal Centre for Remote Sensing (CRTS), various tsunami scenarios were analysed in Rabat, with the support of digital models generated by TIDAL software. The CRTS used this study as a starting point to launch a pilot project assessing the risk of tsunamis in Rabat. It presents hazard maps modelled on scenarios such as the Great Lisbon Earthquake of 1755 and a possible future landslide in La Palma (Canary Islands). These maps incorporate potential waves' maximum height at each point along the coast, as well as potential flood zones. As part of the study, various other indexes were produced such as a flood zone building vulnerability index and maps illustrating the damage levels for each flooded building. Damage levels were calculated by accounting for the modelled wave's height and the type of building under consideration, following a methodology developed by the SCHEMA Project (www.schemaproject.org, financed by the European Commission).

In parallel, the CRTS organised a technical workshop to present and discuss the main findings of its analysis with all the stakeholders involved in natural risk management in Morocco. The various maps produced were also shared with potential users, whose feedback was then integrated in the final versions.

Source: Interviews conducted by the OECD as part of a study on risk management in Morocco (May 2015).

Exposure, at-risk elements and vulnerability

Once natural hazards have been identified, analysed and mapped, this information can be cross-referenced with data relating to population and other at-risk elements along with their vulnerability, in order to obtain an assessment of the potential damage that could be caused by the principal risks. Because this type of information requires a combination of numerous geographical and socioeconomic data sources, ensuring its availability in a standard format is challenging for institutions responsible for risk management in OECD countries.

In Morocco, the High Commission for Planning manages socio-economic, georeferenced datasets and carries out regular surveys on the population and has cultivated accurate information on land use, population, demographic trends and household income. However, very little map-based information is directly accessible online. A number of Geographic Information Systems have been installed within various administrative bodies in Morocco, with a focus on urban agencies that gather a large amount of information concerning exposure in urban environments for town planning documentation. Major infrastructure is also listed by the Ministry of Public Works or other operators responsible for it.

By collaborating with these different institutions, the Ministry of General Affairs and Governance and the World Bank ran models between 2008 and 2012 and developed a database with the main areas of strategic importance for Morocco and their exposure to risks. Previous risk assessment initiatives had focused on basic vulnerability criteria. Four vulnerability categories were established for the assessment of flood-risks, ranging from low to very high, based on the type of exposed land use. Impact criteria for the human, agricultural, industrial and environmental assets were also established (Table 3.2).

Table 3.2. Vulnerability criteria for the flood prevention programme in Morocco

Low vulnerability	Medium vulnerability	High vulnerability	Very high vulnerability
Desert areas **Forests** **Natural areas** **Wildland**	Large villages Extensive farming areas Secondary infrastructure Household waste facilities	Sparsely populated areas Intensive farming areas, irrigated areas, market garden areas, fruit-growing areas and palm groves Areas of minor activity Main infrastructure Non-toxic industrial waste facilities	Heavily inhabited areas Campsites Areas of major activity Very busy tourist areas Toxic industrial waste facilities

Source: Ministry of Water (2003), National Flood Protection Plan.

Data on losses and damage

Data on losses and damage caused by disasters constitute an extremely pertinent source of information for risk assessment. The availability of this information enables a long-term assessment of losses caused by natural disasters and provides an essential benchmark to standardise risk assessment models. Furthermore, these criteria contribute to the assessment of the Sustainable Development Goals advocated by the United Nations in 2015.

Since a number of different actors possess fragmented parts of such information, there is no comprehensive database with extensive records of damage in Morocco. Although the *Desinventar* database was developed specifically for this purpose by the United Nations Office for Disaster Risk Reduction (UNISDR) and installed at the Ministry of the Environment, the database does not seem to be subject to regular updates. A 2015 risk prevention project implemented jointly by the Ministry and UNISDR in the MENA region attempted to update the database, but the information collected remains limited, especially regarding flooding. The Ministry of the Interior's Monitoring and Coordination Centre gathers a large amount of post-flood facts and feedback, often based on information received from *wilayas*. However, the Monitoring and Coordination Centre does not possess a truly integrated database that details losses resulting from major disasters. During the design of the National Flood Protection Programme, the Ministry of Water had also developed a GIS and a database which aimed to record damage caused by floods, but there were too few entries for an effective use of the tool for risk assessment. These examples demonstrate the difficulties involved in sharing risk-related information among administrative bodies in Morocco (Box 3.7).

The mapping and monitoring of flood zones by the Royal Centre for Remote Sensing (CRTS) also presents very relevant sources of information on the areas affected by floods, as well as any spatial or temporal changes observed. Data gathered by the Ministry of Public Works on the different types of damage on infrastructure, by the Ministry of Agriculture on agricultural production, or by the *wilayas*, provinces and *préfectures* during local risk assessment processes constitute vital sources of information for risk modelling and assessment. They help improve the standardisation of digital models and ensure they are as realistic as possible. In fact, the validation of such methods is absolutely essential to develop well-informed public policies.

Box 3.7. National Drought Observatory

As part of the national drought adaptation strategy and the battle against the effects of drought, the former Ministry of Agriculture, Rural Development, Water and Forests established a National Drought Observatory (ONS) in 2001. Its objective was to contribute to the development of a national drought adaptation strategy and to its implementation.

Designed to serve as an institutional-level forum where scientific and technical information could be exchanged, the ONS centralises meteorological, hydrological, agronomic and socio-economic data originating from the different partners involved in drought risk management. Monitoring seasonal effects as well as other long-term parameters allows the ONS to develop several indicators based on reliable information and to contribute to enhanced decision-making processes at various levels of the drought management chain, including at the prevention and crisis management stages.

Since the launch of the initiative, the central management unit has been accommodated within scientific facilities in order to guarantee its neutrality and efficiency during the implementation of ONS' action plans. It is also housed within research facilities, national and regional ministerial departments, and bodies run by various research institutes and universities.

Source: Interviews conducted as part of the OECD Study of Risk Management in Morocco (May 2015), Hassan II Agronomic and Veterinary Institute (IAV Hassan II), *"The National Drought Observatory: A strategic decision support tool"*, newsletter from the Hassan II Agronomic and Veterinary Institute, n°2, March-April 2001, http://www.agriMorocco.net/iavinfo2.pdf.

An incomplete risk assessment process

It is clear the substantial efforts undertaken over the past 20 years in Morocco have led to a clear improvement in risk-related knowledge. Various experts working in scientific and technical institutions, in academia and universities, or in engineering offices, represent key assets to help ensure risk assessment continues to inform risk management policies. In addition, a large amount of data is readily available to effectively contribute towards this process. Such expertise and data have paved the way for advancing critical risk management in Morocco, such as the paraseismic construction regulations, the National Flood Protection Plan and the conceptualisation of an insurance mechanism currently underway to cover individual losses.

However, according to various experts and stakeholders, the risk assessment process in Morocco seems incomplete. Despite the development of efficient but under-utilised tools, several key stakeholders, particularly at the local level, do not have access to accurate information to inform decision-making. It seems, therefore, the efforts have not been replicated everywhere, the tools deployed have not been regularly updated, and their appropriation has remained low. Some deficiencies need to be resolved in Morocco's risk assessment processes, in particular with regards to risk assessment and risk mapping at the local level, where they aim to inform policies relating to town/local planning, risk scenarios development for planning emergency resources, and their corresponding emergency response plans. Generally, the risk assessment process should ensure all the existing measures are consistent and information needs relating to risk are met (Box 3.8).

Box 3.8. OECD/G20 disaster risk assessment framework

During the G20 summit in Los Cabos (Mexico) on 29 June 2012, the Ministers of Finance asked the OECD to develop a reference framework for risk assessment. This framework stresses the following points in particular:

- Adopt a multi-risk and integrated risk assessment approach.
- Clearly define the objectives of risk assessment, agree on the definitions of the principal terms and on the methodology to use.
- Ensure transparency regarding methodology and information sources, and promote accountability in the risk assessment process.
- Identify and include all essential stakeholders in the risk assessment process.
- Nominate an authority to coordinate the risk assessment process, oversee the set-up of consultation mechanisms, and ensure the exchange between scientific institutions, operators of critical infrastructure and local government.
- Encourage close coordination between risk assessments carried out at national level and those conducted at local level.

Source: OECD (2012), "Disaster risk assessment and risk financing", *G20/OECD Methodological Framework on Disaster Risk Assessment and Risk,* OECD Publishing, Paris, http://www.oecd.org/gov/risk/G20disasterriskmanagement.pdf.

Three different and non-exclusive models developed in OECD countries could serve as an example for Morocco:

- Creating an independent risk observatory such as the National Risk Observatory in France (Box 3.9), which convenes data providers and their users, including private companies, regional authorities and scientific institutions.
- Creating a flexible partnership between various scientific and technical institutions, based on the Natural Hazards Partnership model in the United Kingdom (Box 3.10)
- Establishing a risk-focused institution within a ministry such as the Moroccan Ministry for the Environment, the Moroccan Ministry of General Affairs and Governance or the Moroccan Ministry of the Interior, which would be a source of risk-related expertise, based on the CENAPRED model in Mexico (Box 3.11).

Whichever institutional setup, the newly established body would have to be in charge of the risk assessment process in Morocco, collect and disseminate risk-related data and information, develop methodologies and guidelines to ensure risk assessments are carried out at different levels, facilitate the launch of a transparent national dialogue on risks, and develop standards related to acceptable levels of risk,

Box 3.9. National Natural Risk Observatory in France

The National Natural Risk Observatory (ONRN) was jointly established by the French General Directorate for Risk Prevention (DGPR), part of the French Ministry of Ecology, Sustainable Development and Energy (MEDDE), the CCR (French reinsurance company), and private insurance companies, represented by MRN, on 3 May 2012. The Observatory was set up by the Advisory Board for the Prevention of Critical Natural Risks (COPRNM), particularly as a response to the parliamentary information report that followed the floods caused by Cyclone Xynthia in Charente-Maritime department.

As a public-private instrument that shares and disseminates data and indicators on natural risk assessment, the observatory aims to connect data providers and users of risk-related information. The ONRN has five main objectives: improve and harness the potential of existing knowledge of hazards and related issues, supply an assessment and forecasting system, contribute to risk prevention management and governance, support economic analysis of prevention and crisis management, and contribute towards improving the culture of risk. The ONRN therefore has to structure, make accessible and enhance knowledge, and – above all – produce national and regional indicators.

Source: OECD (2014), *OECD Study on the Management of Flood Risks: the Seine in Paris Region 2014*, OECD Publishing, Paris.

Box 3.10. Natural Hazards Partnership in Great Britain

The Natural Hazards Partnership (NHP) was established in 2011 as a consortium of twelve British scientific and technical bodies and five British government partners to ensure that information and analyses on natural risks are available to society as a whole. The NHP's objective is to support the development of policies, communications and services related to risk. It is a forum that enables the exchange of data, information and research about all forms of risk analysis, including meteorology, the environment, geology and Earth observation, and health. Models studying the impact of natural disasters allow the combination of data and partner expertise in order to identify areas and property most vulnerable to a particular risk. Furthermore, the Partnership contributes to the National Risk Assessment (NRA), which draws attention to new risks and submits recommendations based on potential scenarios. A Daily Hazard assessment provides an overview of any foreseeable natural hazards in the upcoming five days, as well as a general outlook for the next 30 days. This is put together though the combination of natural risk-related information received from different aforementioned technical bodies. Since its inception, the Partnership has prompted considerable enhancement of coordination between different parties by avoiding duplication of effort.

Source: OECD (2016), Toolkit on Risk Governance, https://www.oecd.org/governance/toolkit-on-risk-governance/, reviewed 12 February 2015.

Box 3.11. National Centre for Disaster Risk Prevention in Mexico

Opened in 1988 within the Mexican Ministry of the Interior's National Civil Protection System (SINAPROC), CENAPRED (*Centro National de Prevención de Desastres*) conducts research, offers training sessions, disseminates information, and engages in prevention and monitoring of natural and anthropogenic phenomena that can cause disasters.

CENAPRED has played a vital role in the development of shared risk assessment and knowledge at national level. It has built strong links with the scientific community (universities and research institutes), which enabled Mexican authorities to boost their national risk assessments with robust risk-related data, particularly on earthquakes, floods and tropical cyclones. CENAPRED also provides technical support to local authorities to efficiently develop their own risk atlases and to ensure they meet national assessment standards.

Source: OECD (2013), *OECD Reviews of Risk Management Policies: Mexico 2013: Review of the Mexican National Civil Protection System*, OECD Publishing, Paris.

In practice, a risk mapping system should be standardised nationwide. Definitions of geographical units, appropriate scales, a network of references, area classification based on hazard occurrence probability, amplitude indicators for certain parameters; measurements or exposition levels of various assets are all usually part of the expected characteristics of a mapping tool. An integrated system featuring an assessment of potential damage would be more ambitious, but less common. The European Directive on the assessment and management of flood risks (Directive N° 2007/60/CE, Article 6), previously cited in this document, follows this logic and presents types of indicators (extent of flooding, water height or level, current speed and flow rate) and the exposure level of assets (indicative number of inhabitants, type of economic activity, classification of facilities that could possibly be affected).

Conclusion and recommendations

Following the plans and initiatives launched by Morocco as a response to natural disasters, risk assessment has significantly advanced since the start of the 2000s. Multi-risk methods have gradually developed from single-risk approaches.

A large amount of expertise exists, but it remains fragmented. Tools (such as the atlas of risks) and expert bodies (universities, institutes and research centres) are yet to be professionalised. This would involve making their resources more sustainable, harmonising their methodologies wherever possible, further detailing their gradation/scale and number of indicators, ensuring archives are well kept, databases regularly updated, and improving access to their data.

The issue of methodology and timeframe is crucial because natural risks and their associated risks have very different risk profiles (recurrences, localisations, impact). It is therefore essential to be able to compare these profiles to decide on the allocation of resources. The risk of tsunamis, for example, appears extremely remote (i.e. highly improbable in the short term). However, it would cause much more damage to Morocco in the short and medium term than flooding, although it has a more recurrent risk profile.

Local levels of government (regions, provinces, municipalities) still seem to have different understandings of risk, risk mapping and expected damage, resulting in an inconsistent atlas in Morocco. The integration of the damage factor is also still lacking.

Once the expertise and tools have been strengthened, an explicit data policy should guide the access to risk-related information. The observed lack of discussion and coordination among data providers goes hand-in-hand with access issues encountered by the users.

A large part of the information is either of scientific nature (with a research perspective) or of operational nature (a dedicated, proprietary tool, in the case of public agencies). This is why all or part of the databases are not truly accessible to public decision-makers (town-planning agencies, for which certain information must be of a legal nature) or even to private decision-makers (citizens and investors).

Recommendations

ESTABLISH A NATIONAL RISK ASSESSMENT POLICY

Morocco's risk management strategy needs to be grounded on a national assessment. This requirement is based on an OECD Recommendation on the Governance of Risks, which received encouragement from the G20 on 29 June 2012. The idea is to create a framework in Morocco that will serve as a reference for all decisions and policies related to risk management (prevention, emergency management resources, rebuilding, financing, risk culture, and citizens' support).

As this recommendation is highly demanding, it requires sufficient time to be implemented. The establishment of a risk observatory and the mobilisation of necessary resources and methods would be among the top priorities.

Establish a Critical Risk Observatory, or an equivalent body.

A neutral hub should be established (observatory, national centre or institutionalised partnership) to serve as a foundation for national and local risk assessment in a coordinated and prospective manner. As much as possible, this observatory would not just limit itself to natural risks but would also include industrial, technological and anthropomorphic risks. This structure would steer methodological efforts, gather various data, archive and keep them updated, and ensure a smoother exchange of information between different stakeholders.

Combine university and technical expertise and rally experts from Morocco's scientific community through a public research programme to assess risks.

This programme would involve the development of an exhaustive database of losses and damage caused by disasters, mapping of flood extension areas in Morocco for different return periods, seismic micro-zoning of areas that are most at risk from earthquakes, mapping of tsunami risks along the entire Moroccan coast, and observing trend changes, the prospects of locust infestations, as well as industrial risks and classified facilities across the country.

Develop national risk assessment standards.

Implementing the principles (integration, territoriality, prevention) and the mechanisms of Article 8 of the Framework Law on environment N°99-12 would, for example, require hazard, risk and impact studies to be harmonised at national and then territorial level. Consequently, issues regarding methodology and standards cannot be ignored.

Develop risk mapping at territorial level along with updated records and capacity building programmes in tandem with regional authorities.

This would involve the development of an incentive framework for risk assessment at local and regional levels, as well as corresponding methodologies and guidelines. A capacity building programme should also be established for aspects relating to regulatory framework, assessment methods, and hazard/risk knowledge.

Once the necessary expertise and tools have been reinforced, an explicit data policy should guide the access to risk-related information.

The observed lack of discussion and coordination among data providers goes hand-in-hand with access issues encountered by the users. A data policy would also clarify who has access to which types of data, under what conditions, and for what usage.

Bibliography

Oum Er-Rbia Hydraulic Basin Agency (2012), "Summary", Draft master plan for the Integrated Management of Water Resources of the Oum Er-Rbia Basin and Atlantic Coastal Basins, Oum Er-bia Hydraulic Basin Agency, www.abhoer.ma/pages_externes/PDF%20pdaire%20francais/R%C3%A9sum%C3%A9%20d%C3%A9finitif.pdf.

Bouregreg and Chaouia Hydraulic Basin Agency, "Executive Summary", *Draft master plan for the Integrated Management of Water Resources of the Bouregreg and Chaouia Basin*, NOVEC, www.abhatoo.net.ma/content/download/23190/428545/version/1/file/plan_directeur_amenagement_integre_ressources_eau_bouregreg_chaouia.pdf.

World Bank (2014), *Strengthening of resilience in Morocco: input on an integrated risk management strategy*, Washington, D.C.

Benchekroun, S., R. Omira, M.A. Baptista et al. (2013), Tsunami impact and vulnerability in the harbour area of Tangier, Morocco, *Geomatics, Natural Hazards and Risk*, National Centre of Scientific and Technical Research (CNRST), Portuguese Sea and Atmosphere Institute, University of Lisbon, Ibn Tofail University (Kenitra), 2013.

Urban Municipality of Tétouan (2013), Local guide to natural risk prevention in Tétouan.

El Khalki, Y., A. N. Taïbi, A. Benyoucef et al. (2007), "Flood risk management in Tadla-Azilal Region (Morocco): The Beni Mellal example", Act of Congress, 6-9 November 2007, Hanoi, http://www.infotheque.info/fichiers/JSIR-AUF-Hanoi07/articles/AJSIR_2-10_Taibi.pdf.

Hassan II Agronomic and Veterinary Institute (IAV Hassan II), "The National Drought Observatory: A strategic decision support tool", newsletter from the Hassan II Agronomic and Veterinary Institute, n°2, March-April 2001, http://www.agriMorocco.net/iavinfo2.pdf.

Ministry of Water (2003), National Flood Protection Plan.

Pateau, M. (2014), *From hazard to natural risk: The Tangier-Tétouan Region example (Rif, Morocco)*, Geo-Eco-Trop, 2014, vol. 38, no 1, p. 23-32, http://www.geoecotrop.be/uploads/publications/pub_381_03.pdf.

OECD (2012), "Disaster risk assessment and risk financing", *G20/OECD Methodological Framework on Disaster Risk Assessment and Risk Financing*, OECD Publishing, Paris, http://www.oecd.org/gov/risk/G20disasterriskmanagement.pdf.

OECD (2013), *OECD Reviews of Risk Management Policies: Mexico 2013: Review of the Mexican National Civil Protection System*, OECD Publishing, Paris, DOI: http://dx.doi.org/10.1787/9789264192294-en.

OECD (2014), *OECD Study on the Management of Flood Risks: the Seine in Paris Region 2014,* OECD Publishing, Paris.

State Secretariat attached to the Ministry of Energy, Mining, Water and Environment, Water and Environment department (2008a), "Mission 1: Identification of risks – risk of flooding", Study on the mapping of and development of a Geographic Information System for critical risks in Morocco, Études et mesures Les 5 Domaines, http://www.environnement.gov.ma/PDFs/Rapport_Risque-Inondation.pdf.

State Secretariat attached to the Ministry of Energy, Mining, Water and Environment, Water and Environment department (2008b), "Mission 1: Identification of risks – risk of tsunami", Study on the mapping of and development of a Geographic Information System for critical risks in Morocco, Études et mesures Les 5 Domaines, http://www.environnement.gov.ma/PDFs/Rapport_Risque-Tsunami.pdf.

State Secretariat attached to the Ministry of Energy, Mining, Water and Environment, Water and Environment department (2008c), "Mission 1: Identification of risks – risk of earthquakes", Study on the mapping of and development of a Geographic Information System for critical risks in Morocco, Études et mesures Les 5 Domaines, http://www.environnement.gov.ma/PDFs/rapport_risque-sismique.pdf

State Secretariat attached to the Ministry of Energy, Mining, Water and Environment, Water and Environment department, Risk Atlas, http://www.environnement.gov.ma/PDFs/Atlas_Risk.pdf.

Talhaoui, A., A. Iben Brahim, M. Aberkan, et al. (2004), "Seismic microzonation and site effects at Al Hoceima City, Morocco", *Journal of Earthquake Engineering*, Vol. 0, No. 0 (2004) 1-12, Imperial College Press, January 2004.

UNISDR (2015), Morocco: National progress report on the implementation of the Hyogo Framework for Action (2013-2015), http://www.preventionweb.net/files/47166_MAR_NationalHFAprogress_2013-15.pdf.

Werren, G. and M. Lasri (2014), "Practical Guide to Applying the Swiss Method in Morocco", Flood-risk map, Institute of Geography and Sustainability at the University of Lausanne (IGD), Geo-environmental and Planning Test Laboratory at Sidi Mohamed Ben Abdellah University in Fez (LAGEA), https://serval.unil.ch/resource/serval:BIB_D261C5C97A38.P001/REF

Zurich (2015), Moroccan floods in 2014: what can we learn from Guelmim and Sidi Ifni?, www.zurich.com/_/media/dbe/corporate/docs/corporate-responsibility/risk-nexus-morocco-floods-of-2014-november-2015.pdf?la=en

Chapter 4.

Risk prevention in Morocco

Due to external factors such as climate change or internal forces like demographic growth and urbanisation, Morocco finds itself increasingly exposed to critical risks, especially those of natural origin. Improving the country's resilience when confronted with such risks can only be achieved through wide-ranging preventive measures. This chapter assesses these structural and non-structural measures. These involve risk culture, town and land-use planning, improving the resilience of critical networks and ensuring the continuity of business activity, as well as managing hazards through protection infrastructure. Seeking synergy between different actors combined with preventive measures will help strengthen Moroccan society's resilience in the event of a critical risk.

Introduction

A shared risk assessment within an integrated national strategy framework is necessary to make informed decisions and sensible resource allocations throughout the risk management cycle, especially regarding prevention. This allocation should be followed by the design and implementation of more detailed policies regarding structural and non-structural measures, pursuing resilience, financing, and incentivising non-public actors to shoulder some of the prevention burden.

This chapter deals with public policies implemented in Morocco regarding the prevention and mitigation of the risk of natural disasters. It focuses on both structural measures (related to engineering or civil engineering, such as dams or seawalls), and non-structural measures (related to land-use policies, building codes, resettlement programmes or risk awareness) aimed at preventing risks from affecting built up-areas, and reducing the intensity, frequency and impact of potential hazards. The objective is to assess the commitment of Moroccan authorities to develop better prevention mechanisms and the communities' incentives to protect themselves from those risks.

The OECD Recommendation on the Governance of Critical Risks invites governments to mobilise their citizens and businesses around the issue of prevention by increasing their awareness of critical risks and by providing them with incentives to invest in risk prevention and mitigation. To this end, the Recommendation promotes a society-wide approach to risk communication and information based on a dialogue between public authorities and stakeholders, and on incentives that encourage self-protection and resilience. It also calls for consolidating structural and non-structural measures by boosting investment in prevention, including at a local level, and by designing a strategic plan that integrates risk factors. Finally, the Recommendation encourages businesses, especially SMEs, and operators of critical infrastructure (networks in particular) to take steps to ensure the continuity of their activity by developing tools that enable them to better deal with risks.

In Morocco, risk prevention policies were introduced relatively recently, and they are not yet sufficiently supported by an integrated framework for action, or by dedicated tools. Consequently, the planning and prioritisation of prevention efforts remain difficult. Nevertheless, they have benefited from substantial funding, as demonstrated by the major investments made that were part of structural prevention measures introduced in recent years. In the context of economic, social and environmental developments of strategic importance, strengthening prevention policies is key to improving Morocco's risk management policies.

Strengthening risk culture

Strengthening the risk culture among public actors, civil society and the private sector is key to achieving prevention policies. Indeed, the awareness of risks to which citizens, businesses and decision-makers are exposed enables risk-reducing actions at all levels to start taking place. This required the development of risk communication tools that enable everyone to objectively perceive the seriousness and frequency of such risks, as well as the potential impacts on their home, business, town, and way of life or well-being. Stimulated by strong public demand, an improved awareness of risk among citizens can also lead to the emergence and increased support in favour of dynamic government actions in this area.

Risk awareness and memory

In Morocco, despite the recurrent occurrence of natural disasters over recent years, risk awareness does not seem to be aligned to the gravity of the situation at hand. In fact, 88% of the respondents to a survey conducted for this study (Appendices D and E) believed general risk awareness was poor and that people continued to display inappropriate behaviour during natural disasters, which demonstrates a lack of relevant knowledge at local level. As an example, during the fatal Guelmim floods of November and December 2014, communities sometimes recklessly crossed flooded concrete aprons, as they failed to comprehend the extent of the flooding and assumed it was perfectly safe (Zurich, 2015). This significantly increased the death toll and demonstrates the importance of strengthening risk culture in Morocco and raising public awareness of relevant risks.

Morocco also suffers from a gradual loss of collective risk memory due to the long return period of disasters and because their impact is often localised.

In rural areas, traditional flood management know-how is still applied; during the Gharb floods of 2009, some *douars* (traditional village) protected themselves with earthen dykes before the waters reached their highest level. This local knowledge was also demonstrated during the Ourika floods in 1995, when almost all of the numerous victims were Moroccan or foreign tourists rather than inhabitants of the valley. The Guelmim floods affected newly urbanised areas, where traditional irrigation techniques slowing the flow of floodwater with a system of dykes are no longer in use.

While the Al Hoceima earthquake of 2004 remains at the forefront of Morocco's collective memory, it is essential for this memory to be transmitted to the Agadir area, where the last major earthquake took place in 1960. Given the demographic boom experienced by Agadir since then, only a small part of the population can still directly relate to the memory of this major event. Initiatives such as regular commemorations organised by city authorities and the remarkable work conducted by a society of former life sciences professors to raise awareness among school pupils (Box 4.1) are two rare examples of collective risk memory transmission in Morocco.

Box 4.1. Maintaining the memory of the Agadir earthquake

On 29 February 1960, Agadir was shaken by an earthquake of 5.7 on the Richter scale, which resulted in the death of over 12,000 residents – a third of the city's population. The earthquake was the most deadly in Moroccan history and left its mark on several generations. In order to perpetuate the memory of this event, a number of university professors and senior figures within secondary education committed themselves to raising awareness of the risk of earthquakes.

Consequently, Rabat's Mohammed V University, in partnership with Ibn Zohr University in Agadir, organises a day of commemoration attended by local politicians, civil defence representatives and other actors involved in risk management. During this event, the professors raise awareness among local communities about the risk of earthquakes, and share information on recently developed mapping and seismology techniques. They also host discussions with professionals from the construction industry who present the RPS 2000 paraseismic code and answer any questions.

In parallel, some secondary school natural sciences teachers have also joined efforts to raise awareness of this risk and perpetuate the memory of the Agadir earthquake, through the Souss Natural Sciences Association (ASVTS). Established in 2002 and based in Agadir, this organisation runs an awareness-raising project using the memory of the seismic event to create a better risk culture. Among their initiatives, the ASVTS recently held a competition for local schools to create a short film about seismic risk in Agadir. The winning team won a visit to the National Geophysical Institute (ING) in Rabat. In addition, the ASVTS installed a seismic station and an exhibition of the 1960 earthquake to raise awareness of the risk among school pupils, students, parents and educational staff. The acquisition of the station was financed by the National Human Development Initiative (INDH), to the tune of 10,000 Dirhams. Other partners, including the Souss Mass Drâa Regional Council and the Agadir Faculty of Science, have also contributed to funding activities run by the ASVTS. In order to boost the impact of this initiative and reach as many people as possible, the ASVTS plans to move the station to local community centres, certain socio-cultural facilities and various associations for a few days on a regular basis.

Moreover, to mark the 50th anniversary of the Agadir earthquake in 2010, the Moroccan Red Crescent staged a series of commemorative activities, including a photographic exhibition retracing the reconstruction of the Al Inbiat estate, the screening of a documentary film entitled *Children of the Agadir earthquake: 50 years later*, produced by Hassan Bouharouti, and the airing of extracts from the speech delivered by His Majesty Mohammed V the day following the tragedy.

Source: Interviews conducted as part of the OECD Study on Risk Management in Morocco (May 2015); the Souss Natural Sciences Association (ASVTS) (2014), "اللقاء التواصلي لجمعية علوم الحياة والأرض بسوس لموسم 2013/2014" (Meeting of the Souss Natural Sciences Association for the 2013/2014 season), presentation during interviews conducted as part of the OECD Study on Risk Management in Morocco, June 2015; Souss Natural Sciences Association (ASVTS) (2015), "Project: Princess Lala Aicha 'AGLA' educational seismic station", presentation during interviews conducted as part of the OECD Study on Risk Management in Morocco, June 2015.

Public action to raise awareness of risks

The poor level of risk-related knowledge and the disappearance of collective risk memory have amplified the need for ambitious public action in this domain. However, the development of risk culture in Morocco has to overcome the lack of a relevant regulatory framework; unlike in other developed and emerging countries (Box 4.2), it is not compulsory to inform people of risks in Morocco. Thus, while the country's public administration has remarkably advanced its awareness of the importance of risks in recent years, implementing tools and instruments to strengthen this awareness on the ground has been limited.

Box 4.2. Risk information in France and Great Britain

In France, municipalities are required to inform their citizens about risks that may threaten them within the municipality's limits. They must produce Municipal Information Notices on Critical Risks (DICRIM). Their aim is to inform the local population about the risks and what appropriate measures to take. For the public to be made aware of the DICRIM, it must be displayed on a notice board at the town hall and public information meetings must be held biennially. It must be freely accessible and contain all historical risk-related information, and explain all the preventive, protective and safeguarding measures.

In Great Britain, the National Risk Assessment has given rise to local assessments that must be published and made available to residents in a Community Risk Register. For example, assessors in London identified and analysed nearly 60 risk scenarios, which allowed them to set priority activities to improve resilience. London's Community Risk Register includes worst-case scenarios, which are based on scientific and historical data, models and expert judgments.

Source: OECD (2012), "Disaster risk assessment and risk financing", *G20/OECD Methodological Framework on Disaster Risk Assessment and Risk Financing*, OECD Publishing, Paris, http://www.oecd.org/gov/risk/G20disasterriskmanagement.pdf.

While some national public administrations engage in effective risk communication, which illustrates the importance of such an approach, it is not directly oriented towards high-risk populations. The Directorate General of Civil Defence organises an annual open-house initiative in its compound during National Civil Defence Days (Box 4.3). Every summer, the HCEFLCD organises a communication campaign on the risk of forest fires, which includes broadcasting awareness-raising TV and radio spots, distributing posters and leaflets in public places and forested regions, and publishing articles on the topic in the written press, drawing on a budget of 10 to 12 million Dirhams (HCEFLCD, 2015). Meanwhile, the Ministry of National Education has published brochures for schoolchildren (e.g. "Nadir and the floods", "Nadir and the earthquakes") and distributed them in schools. Finally, the Ministry for the Environment has also taken part in these initiatives, organising activities for the International Day for Disaster Reduction, including awareness raising workshops for local elected officials.

Box 4.3. National Civil Defence Day in Morocco

Every year on March 1, Morocco celebrates the National Civil Defence Day. This event is notable for the organisation of various risk awareness initiatives across the country led by the Directorate-General of Civil Defence. These usually take the form of open days and are run by the regional, prefectural and provincial civil defence commands.

In particular, these open days are organised to show visitors the different risks that affect the region in which they live. Awareness and information sessions for local communities are held and leaflets are distributed. Demonstrations of emergency manoeuvres, rescue operations and fire extinguishing drills are also staged. Finally, emergency and rescue equipment is displayed by the DGPC.

The International Day for Disaster Reduction was established by a 1990 decision of the General Assembly of the International Civil Defence Organisation. It aims to draw the attention of the public to the importance of civil defence and to raise awareness of preparatory and self-preventive measures for combating accidents and disasters. It also contributes to paying tribute to the work carried out by national services responsible for combating disasters.

Source: Interviews conducted as part of the OECD Study on Risk Management in Morocco (May 2015). International Civil Defence Organisation (2015), ICDO website, http://www.icdo.org/fr/, reviewed 2 November 2015; Le Matin (2014), "Morocco marks International Day for Disaster Reduction", Le Matin, 28 February 2014, http://www.lematin.ma/journal/-/197687.html, reviewed 2 November 2015.

At the local level, initiatives that raise awareness of risk are limited due to the low risk-related information available. No websites exist to provide direct information about the risks people are exposed to, and which preventive measures should be adopted. Provincial flood prevention commissions organise annual meetings in certain provinces to inform local elected officials of the risk of flooding. Some regional authorities have undertaken exemplary actions, such as in Tétouan, where they have designed a "Local guide to natural risk prevention" and produced leaflets and posters for each risk affecting the region. However, there is generally a lack of well-established, government-approved information about risks, which makes accurate risk communication campaigns difficult. This is partly the cause behind the low awareness levels of public decision-makers and communities at local level. There is also a certain reluctance to engage in risk-related communication with citizens and in specific places (e.g. tourist sites). This points to the need of pedagogical efforts on the benefits of risk-related communication.

Approaches to risk communication

Effective risk communication should combine different communication channels and media to ensure all decision-makers, communities, businesses and professionals are informed about their responsibilities with regards to the risks they are exposed to or which they could help reduce. This can include communication campaigns, site-specific information signs or notices, launching risk-focused websites, supporting civil society organisations, and organising events and exhibitions. Increasingly, OECD countries are also using social networks or smartphone applications dedicated to risk communication (OECD, 2016). To ensure the effectiveness of awareness and culture of risk measures, it is also important to be able to assess their impact. To this end, surveys exploring people's perception of risks can be highly effective tools.

Experts and stakeholders note the limited number of concrete risk-related communication initiatives implemented in Morocco. Consequently, there are no signs or physical markings about risks in high-risk sites, which is a widespread practice in the majority of OECD countries. This can involve placing flood markers indicating a high water mark, or marking off the distance reached inland by an historic tsunami, as is commonly done in Japan, for example. The large rocks displaced by the tsunami of 1755 in the Rabat area could be used as markers for that event, especially during World Tsunami Awareness Day, established on 5 November by the United Nations at the request of Japan. These measures have the advantage of being significantly less costly than others.

Moreover, comments from experts and stakeholders have revealed the need for a more systematic approach to raise awareness in the private sector. Despite requests for more information, from businesses that sometimes choose to commit to internal risk management processes, the information made available to them seems insufficient. Professional associations such as the CGEM can be useful in relaying messages about raising awareness among their members. This was, for example, demonstrated by the Paris Region Chamber of Commerce and Industry, which surveyed Parisian businesses about the risk of flooding from the Seine River. Similarly, in the Loire Basin, targeted information is regularly sent to leaders of businesses located in flood zones. This has led to significant improvement in their level of risk awareness and incentivised them to contribute financially to flood prevention efforts. It is equally crucial to strengthen risk culture among certain professionals who play an essential role in the sphere of resilience through their involvement in town planning, construction or tourism.

Risk-related communication approaches that specifically target vulnerable parts of the population (e.g. tourists, who are generally poorly informed about risks) could be developed. However, this suggestion has met resistance on the ground. The idea of designing leaflets to raise awareness of the risk of floods and distributing them to tourists in the Ourika Valley has encountered reservations from tourist-dependant businesses, which do not wish to "scare off" their customers. Furthermore, there is no specific, systematic approach to raise awareness of the seismic risk in Agadir's hotels, the country's most popular tourist destination. However, the hotel industry is already widely disseminating information to tourists about sustainability-related issues via notices in their rooms (e.g. by encouraging them to be economical in their water usage). By 2030-2050, the number of vulnerable people (tourists, immigrants or the elderly) will grow significantly, and a lack of preventive information in the event of a disaster would be immediately followed by a withdrawal of investors and tourists, who will prefer to go elsewhere. Again, the costs involved in these measures are relatively low.

Importance of education and training in risk management

Strengthening risk culture can also be achieved through risk-related education, i.e. including the subject in school and university curriculums and in professional training sessions. Although the awareness-raising projects run by the Ministry of National Education in schools are worth highlighting, schools should engage in more wide-ranging approaches. In Japan, topics related to risk management are addressed in different subjects throughout students' educational journeys (Box 4.4).

Box 4.4. Risk-related education in Japan

Taking the appropriate measures in the event of an earthquake is essential to save lives. Given its high exposure to the risk of earthquakes, Japan decided to develop awareness programmes that children participate in from an early age, particularly in educational establishments. Past experience has shown young people's awareness of risks can be heightened if they are taught practical skills on a regular basis. To this end, a variety of educational tools are recommended, such as history books, cartoons, games and websites. The goal of these programmes is to improve the ability of students of all ages to actively contribute to risk-reduction efforts in their communities.

For this purpose, the Japanese Ministry of Education provides teaching guidelines. In 2002, new school programmes were introduced to make cross-curricular risk courses compulsory. In 2006, the Ministry adopted a new strategy that incorporated risk awareness into all levels of education. Local approaches taking into account the special features of each risk are encouraged.

Today, all Japanese pupils, from nursery school to high school, attend regular lessons and receive training on risks during the school year. The measure's effectiveness has been demonstrated on various occasions, including during the 2011 tsunami, when all 1,000 pupils attending schools in Kamaishi reacted in a fast, appropriate manner and got themselves to safe zone, despite the massive wave killing over 1,000 residents of the town.

Source: OECD (2016), Toolkit on Risk Governance, https://www.oecd.org/governance/toolkit-on-risk-governance/, consulted on 12 February 2015.

The development of scientific research as well as university and training courses dealing with the issue of risk are also powerful means to raise awareness. Chapter 2 highlighted the quality of research carried out by academia and universities in Morocco. Despite these assets, the link between this type of work and its tangible use in risk management is often weak, which demonstrates the inadequate communication between scientific researchers and public decision-makers.

Additionally, Morocco does not yet offer a university course or a master's degree that truly focuses on risk management, such as those existing in other countries. This kind of education and training could produce graduates that are well trained to hold positions in town planning, in land-use management, and in risk prevention. Such a programme could also contribute to build capacity of Morocco's authorities by designing courses for administration officials, and particularly at the local level, where there is a strong need.

Structural prevention measures

Until now, structural prevention measures have been the favoured approach to risk prevention in Morocco. This is particularly the case for the fight against flooding and drought, which has led to the development of hydraulic infrastructure. The policy of large dams launched in the 1960s, the National Flood Protection Plan (launched in 2002), and more recently, the north-south water transfer, constitute the three main pillars of this approach.

Prevalence of structural measures

Dams policy in Morocco

Since the 1960s, Morocco introduced a water policy that was based on the construction of major dams in order to ensure the mobilisation of the country's water resources for the development of irrigation and other uses. Several large dams were built in response to significant floods: the Hassan Addakhil Dam was constructed after the devastating Ziz Valley flood in November 1965, the Mohammed V Dam was reinforced following the Moulouya flood in May 1963, and the Boukarkour and Hassar Dams were built after the Mohammedia flood in 2002 (Table 4.1).

This large dam policy remains dominant today: in 2015, the construction of seven new major dams, amounting to 7bn Dirhams (El Ghomari, 2015), was scheduled. By 2030, the National Water Plan, an extension of the National Water Strategy, expects investments of 28bn Dirhams for the construction of large dams, part of which should contribute to flood-risk prevention.

The development of major infrastructure has, therefore, limited worst-case scenarios in a number of river basins. However, questions are now emerging about whether this policy should be pursued given the rising cost of new building works caused by the reduced availability of construction sites, uncertainty linked to the effects of climate change, and their high maintenance costs.

Table 4.1. Principal dams in Morocco playing a protective role against flooding

Name of dam	Province	Date opened	Main function	Height (in metres)	Capacity (in mm³)	Protected area
Mellah	Ben Slimane	1931	I, SDIW	33	8.8	Mohammedia
El Kansera	Khemisset	1935	E, I, SDIW	68	266	Douar Oued Beht
Mohammed V	Oujda	1967	E, I, SDIW	64	410	Area downstream from the dam on the Moulouya River
Hassan Addakhil	Errachidia	1971	I, PF	85	347	Ziz Valley
Mansour Eddahbi	Ouarzazate	1972	E, I, FC	70	529	Draa Valley
Idriss Ier	Taounate	1973	E,I	72	1186	Gharb Plain
Sidi Med Ben Abdallah	Rabat	1974	SDIW	99	486	Bouregreg Valley
Oued El Makhazine	Larache	1979	E, I, SDIW	67	773	El Ksar El Kebir and Loukkos Plain
Sfa	Agadir	1985	FC	17	0.6	Conglomerations bordering *Oued* Sfa
Tizguit Aval	Ifrane	1986	FC	18	0.12	Areas downstream of Tizguit River
Azib Douirani	Marrakesh	1987	I, PF, FC	16	0.6	Oulad Brahim Village
Aman Seyernine	Meknes	1987	I, FC	16	0.35	Boufekrane Village
Mouillah	Khouribga	1987	I, FC	16	0.5	Boujad
Imaouene	Guelmim	1992	FC	23	0.23	Bouizakrane Centre
Agherghis	Guelmim	1992	FC	24	0.28	Bouizakrane Centre

Table 4.1. Principal dams in Morocco playing a protective role against flooding *(continued)*

Name of dam	Province	Date opened	Main function	Height (in metres)	Capacity (in mm³)	Protected area
Aggay	Sefrou	1994	FC	40	1.25	Sefrou
Saquia El Hamra	Laayoune	1995	FC, EN	16	110	Laayoune-Tarfaya Road
Al Wahda	Sidi Kacem	1997	I, E, FC	88	3800	Gharb Plain
Ben Yekhlef	Berkane	1997	FC	18	0.03	Berkane
El Menzel	Berkane	1998	FC	18	0.16	Berkane
Hassan II	Taourirt	2000	I, SDIW, FC	91	275	Oued Za Valley
Ahmed El Hanssali	Beni-Mellal	2001	I, SDIW, FC	101	740	Areas downstream of Oum Er-Rbia River
Reg Aouin Kora	Tan Tan	2005	FC, WL, GR	16	1.2	U
Kheng Messaoud	Tan Tan	2005	FC, WL, GR	24	1.3	U
Hassar	Mohammedia	2005	FC	15	2.13	U
Touizgui Remz	Assa Zag	2007	WL, I, GR, FC	18	78	U
Al Himer	Settat	2008	FC, GR	38.5	14	U
Koudiat El Garn	Settat	In progress	FC	54.5	33	U
Sidi El Mahjoub	-	In progress	WL, I, FC	22	1	U

Notes: Legend: SDIW: Supply of Drinking and Industrial Water; E: Energy; I: Irrigation; PF: Protection against Floods; FC: Flood control; GR: Groundwater Recharge; WL: Watering Livestock; U: unknown.

Source: Ministry of Water (2015).

National Flood Protection Plan (PNPI)

Local protection measures are part of a 2002 investment programme aimed at protecting priority sites exposed to a once-a-century risk, in line with international standards. Developed by the Ministry of Water and using a hazard assessment study, the National Flood Protection Plan identified 391 endangered sites nationwide, including 50 priority locations (Chapter 2). The programme considered structural flood-control measures such as dykes, hydraulic weirs, hill dams, protective walls, and embankment protection work. The total amount of investment required by 2017 was estimated at 25bn Dirhams. The programme also included non-structural measures, specifically the enhancement of early warning systems, covered in Chapter 5. The Hydraulic Basin Agencies were assigned the responsibility of implementing these measures and adjusting them according to their Integrated Management Master Plans for Water Resources.

Since then, numerous studies have been conducted on various sites and work has been carried out to the tune of 7.6bn Dirhams, representing a major investment in risk prevention in Morocco (Table 4.2). In total, almost a third of the estimated financing was raised, 70% of which was covered by the Ministry of Water. Initially, these targeted the most easily identifiable priority sites (Mohammedia, Berrechid, Settat, and Casablanca

Mohammed V International Airport), but a total of 244 sites were the subject of studies or construction work. In 2009, the Moroccan Audit Office noted the lack of a flood management strategy among the Hydraulic Basin Agencies, and the low completion rate of work compared to the number of studies that had been conducted (Audit Office, 2009). However, programme implementation has progressed since then, reaching a current rate of 15 to 20 sites per year. The funding mobilised to date remains below the initial estimates, while the completion of the programme depends on additional major financing.

Table 4.2. Completed PNPI projects between 2003 and 2015

	Initial plan				Work completed			
Hydraulic Basin Agencies	Problem spots listed		Cost of planned work (priority sites)		Problem spots dealt with		Cost of work undertaken	
	Number	% of total	Dirhams (millions)	% of total	Number	% of total	Dirhams (millions)	% of total
Bouregreg Chaouia	13	3	356	7	30	12	2,750	39
Loukkos	40	10	1,758	32	29	12	1,048	15
Moulouya	62	16	434	8	22	9	688	10
Oum Er-Rbia	58	15	410	8	34	14	348	5
Sebou	52	13	578	11	73	30	1,504	21
Souss Massa Draa	99	25	1,157	21	29	12	967	14
Tensift	50	13	760	14	16	7	231	3
Ziz Guir Ghriss	16	4	-	-	10	4	26	0
TOTAL	390	100	5,453	100	244	100	7,135	100

Source: Ministry of Water

There has also been a noticeable change in the order of priorities from the initial plan to the actual work carried out, with investment focused on the Bouregreg (39% of the investments instead of the 7% initially envisaged, and 30 problem spots resolved instead of the 13 initially identified) as well as the Sebou, to the detriment of the Loukkos and Souss-Massa-Dra basins. This was a consequence of the major flooding that took place in 2002 in Casablanca Region, and the recurrent flooding in the Gharb basin in 2008, 2009 and 2010. This disparity presents challenges for the decision-making process in terms of the prioritisation of prevention work. The update to this programme announced by the Ministry of Water will provide an opportunity to better clarify the choices made, review the corresponding financing models and take into account changes to risk factors over the past 10 years.

The prevalence of structural measures, such as those promoted by the PNPI, has revealed limitations during recent events. For example, the construction of the Assif El Krayma hill dam in 2012 did not prevent a major flood from reaching Sidi Ifni in November 2014. Although the dam did initially limit the impact of Oued Ifni flood, it was still overcome by the scale of the phenomenon and failed to fulfil its protective function (Zurich, 2015). It is, therefore, important to promote complementary measures, particularly non-structural ones, to compensate for any potential inadequacies.

North-south water transfer project

The Ministry of Water also plans to oversee a major water transfer project, from the more rain-soaked north to the south of the country. This project is an integral part of the National Water Strategy and it involves transferring 845mm^3 of water per year over time. This represents 5% of Morocco's average annual flows, covering a distance of 450km, between the Loukkos and Sebou basins to the water-stressed Oum Er-Rbia, Tensift and Bouregreg basins (Ministry of Water, 2015). The objective is to enable water management in an organised manner at national level and to reduce the rainfall imbalance that exists across the kingdom's different regions. More specifically, the transfer will help ensure a stable water supply for the Marrakesh Region and the coastal areas between Rabat and Casablanca, to recharge and preserve groundwater in the critically over-exploited Haouz basin, to curb the water deficit in the irrigated zones of Doukkala Region, and to conserve the coastal aquifers from Rabat to Safi and in Berrechid, as these are necessary for farming (El Ghomari, 2015).

Completing such a development project is a major challenge for Morocco. Its cost, estimated at 30bn Dirhams, will represent a substantial investment in terms of national budget. The impact of climate change on water resources is concerning, but the project could lead to significant changes in regional balance that would require an accurate analysis of long-term needs for different areas that are affected by the water transfer. This initiative illustrates Morocco's need to address the question of structural risk prevention infrastructure in the context of an informed and transparent debate and a well-established governance framework. This project could represent such an opportunity, as it is highly likely to stimulate people and bring them together.

Governance of structural protection infrastructure

While the approach of developing structural risk prevention measures is preferred in a number of countries, these policies should go hand-in-hand with well-established governance mechanisms. The high costs of these types of projects and their pressure on public finances raise questions regarding the optimisation of decision-making mechanisms that lead to these choices.

Decision-making mechanisms

In Morocco, the development of protection infrastructure is managed centrally, focused on sector-based approaches, and essentially funded by the national budget. The Ministry of Water has developed, managed, funded and implemented these policies, with some funding from international donors for dam projects. While this process has occasionally been adapted to local conditions, especially for the development of ABH action programmes or to regions, where provincial flood-prevention commissions have been quite active, these major programmes rarely connect various stakeholders with their decision-making mechanisms. The north-south transfer project is typical of such a centralised and top-down approach that requires raising significant amounts for funding.

Today, as it embarks on a journey to implement such an infrastructure-related policy, Morocco faces numerous challenges. The cost of dams is increasing due to the reduced availability of suitable sites, while the first constructions were completed on sites yielding the greatest knock-on benefits. The PNPI also initially favoured sites suffering from the greatest vulnerability to flooding or those that had just been affected by a significant flood, but the recurrence of these events and their unprecedented scale in recent years contributed to making such decisions difficult to reach. In addition, uncertainty linked to the effects of climate change and its impact on water resources and extreme events further complicates the situation. The fulfilment of these programmes also requires finding additional funding to supplement the insufficient amount provided by both the Ministry of Water and the Hydraulic Basin Agencies. This, therefore, requires the involvement of additional stakeholders in the decision-making process and general agreement on criteria to prioritise projects.

In most developed and emerging countries, decision-making mechanisms on risk prevention have been designed to convene stakeholders, acquire financing from different levels of government and the private sector via incentive-based approaches, and assess and compare various reduction options to select the most effective approach based on an economic costs-benefits analysis. Concerning infrastructure, various pertinent questions continue to arise on the social and environmental impact and the inclusion of maintenance and operation issues.

This approach has started to emerge recently in Morocco, with the launch of several initiatives. For example, protective activities were carried out in Casablanca after recurrent flooding from *oued* Bouskoura (on which the city is built), which prompted the creation of a broad partnership between the state, regional authorities and private businesses to fund the "Super Collector West" water redirection project (Box 4.5). The call for tenders initiated by the FLCN in 2015 is another example of such an approach. Open to all national administrations and local authorities, and based on transparent assessment criteria, the FLCN will jointly fund 23 prevention projects – mainly involving structural measures run by regional authorities (more details on this programme are available in the financing section). Partnership-based approaches are expected to continue emerging, particularly in cooperation with municipalities as these are also responsible for flood defence infrastructure development. Concerning the environmental and social impact, the application of Law n°12-03 entails the development of impact studies for infrastructure projects that should be accompanied by corrective measures approved by regional committees responsible for these issues.

Box 4.5. Western Casablanca Super Collector: public-private financing

Launched in 2014, the "Super Collector West" project aims to equip Casablanca by March 2017 with a collection and protection facility that will redirect floodwater from Oued Bouskoura away from the city and towards the sea. The project gained the financial backing of a variety of actors, both public and private. Partly built on the natural bed of Oued Bouskoura, the city of Casablanca is frequently threatened by flooding. In 2010, it was severely affected by flooding that caused considerable damage all along the *oued*'s original route. With 3.36 million residents and representing 23.4% of Morocco's GDP, it was essential for Casablanca to develop infrastructure providing long-lasting protection from the risk of flood.

Casablanca's Super Collector West project envisages counteracting floods with a return period of 100 years. Managed by Casa Aménagement– a company implementing major development projects in line with the Greater Casablanca Master Plan for Urban Development –, the project involves the creation of reservoir storage in the catchment basin, the diversion of Oued Bouskoura towards Oued Merzeg, and the extension of the tunnelled waterway across the city. The construction was divided into two distinct phases: first, an open 3km-long drainage canal with variable slopes and sections adapted to the natural morphology of the *oued*, and second, a gallery consisting of an underground tunnel and backfilled trench, 6.2km in length and 5.5m in diameter, that runs out to the sea.

This large-scale project required a total budget of 855 million Dirhams. Given the substantial financial needs involved, it was decided to set up an innovative financial partnership, enlisting various public and private actors involved in flood prevention in Casablanca. The overall cost of the project was distributed in the following manner:

- General state budget: 120 million Dirhams.
- Fund to Combat the Effects of Natural Disasters: 120 million Dirhams.
- Ministry of the Interior (DGCL): 150 million Dirhams.
- Ministry of Energy, Mining, Water and Environment: 75 million Dirhams.
- Urban Municipality of Casablanca (FDT): 150 million Dirhams.
- Casablanca-Settat Region: 100 million Dirhams.
- Office Chérifien des phosphates (OCP): 80 million Dirhams.
- Anfa Town Planning and Development Agency (AUDA): 50 million Dirhams.
- Bouregreg and Chaouia Hydraulic Basin Agencies: 10 million Dirhams.

Furthermore, this participatory public-private financing mechanism was consolidated by an institutional partnership: a technical committee and a monitoring committee are scheduled, while their respective modes of operation, roles and commitments have already been defined.

Sources: Interviews conducted as part of the OECD Study on Risk Management in Morocco (May 2015) ; Casa Aménagement (2014), Super Collector West website, http://www.casa-amenagement.ma/fr/nos-projets/super-collecteur-ouest, reviewed 21 December 2015; Michel-Kerjan, Erwann O. (2014), *CAS-FLCN: Information for defining priority areas and selection criteria for projects requiring funding and implementation*, report produced as part of the World Bank's support work for the Ministry of the Interior.

Operations and maintenance

Good infrastructure governance also requires careful consideration of questions relating to long-term operation and maintenance, and preferably since the design stage.

This issue is particularly striking in Morocco, given the high erosion levels caused by the geological conditions and climate of the country. Steep slopes and stormy rains falling on dry soils encourage erosion, which can result in major sedimentation of rivers and engineering structures. Consequently, 10% of Moroccan dams are now filled with sediment, and these could represent 39% of the dams located in the Moulouya basin. Some dams are already completely silted up. Although techniques do exist to resolve this problem, such as dredging, their cost is often prohibitive (Lemaizi, 2015). The effectiveness of these structures is subsequently questioned if they can no longer perform the role for which they were built.

The problems encountered with dams highlight the importance of infrastructure maintenance. Such efforts should, in principle, be planned and funded through a regular budget, and implemented by skilled labourers to ensure the costly infrastructure can have the longest possible life-span, and consequently improve its economic efficiency. Local authorities are responsible for cleaning and waterway upkeep, as well as the maintenance of minor hydraulic infrastructure, but they do not always have sufficient budgets dedicated to financing such tasks. Furthermore, the costs involved can often be extremely high. For example, the annual cost of cleaning the entire Gharb Valley sanitation system has been estimated at 280 million Dirhams. In the absence of an accessible database listing all infrastructures and its corresponding condition to serve as a basis for planning maintenance, it is difficult to obtain an estimate of the state of protective infrastructure in Morocco.

Ensuring clarity and flexibility while establishing operational rules for protective infrastructure can also present a challenge to achieving good risk governance. There can sometimes be competition between various usages of protective infrastructure. The effects of climate change and evolving usages call for regularly reviewing the rules to guarantee their alignment with risk factors changes. Several international examples have demonstrated that poor infrastructure governance can have dramatic repercussions, such as during the flooding of Brisbane, Australia (Box 4.6). Morocco's flexibility in modifying the operating rules of a major dam located in the upstream section of the Sebou River constitutes a good model in this respect. After the demonstration that the economic impact of repeated flooding in the Gharb Plain far exceeded the economic losses caused by a reduction in electricity production, the National Water and Electricity Authority operating the dam agreed to reduce the maximum water level prior to the rainy season.

Box 4.6. Operational effectiveness of protective structures in times of crisis in Australia

The 2011 floods in Queensland, in Brisbane in particular, demonstrated the importance of guaranteeing the operational effectiveness of protective structures when confronted with the risk of flooding, as well as regularly reviewing their conditions of use. The hydrologists tasked with investigating the damage caused by the floods in Brisbane, Ipswich, Toowoomba and Lockyer Valley found massive releases of water from the saturated Wivenhoe Dam had been a determining factor in the downstream flooding of 11 and 12 January 2011. In their opinion, the dam's operational inefficiency contributed significantly to the extent of flooding in Brisbane and the surrounding area, which caused major damage. Despite warnings from the Australian Bureau of Meteorology about the potential force of La Niña, operation managers at Wivenhoe Dam did not activate any drainage of the reservoir because the structure's operating manual did not include the implementation of such a strategy. Regularly reassessing the operating conditions of protective structures is, therefore, essential to guaranteeing their operational effectiveness.

Source: Queensland Flood Commission of Inquiry (2012), Final report, http://www.floodcommission.qld.gov.au/__data/assets/pdf_file/0007/11698/QFCI-Final-Report-March-2012.pdf.

Non-structural prevention measures

According to the majority of local and national stakeholders encountered, the key issues with regards to risk prevention are spatial planning, land use and town planning. The growing level of urbanisation creates new risks, particularly in relation to informal, substandard housing and the lack of basic infrastructure. To limit vulnerable communities' exposure to risk, this urban development must be accompanied by appropriate land-use and town planning policies. In addition, the ongoing development of infrastructure offers new opportunities to further strengthen the risk prevention process.

Difficulties of implementing town planning policies that incorporate risk

Existing urban planning documents do not address the question of risk beyond the provision of basic information. However, municipalities grant building permits based on these documents. The Master Plans for Urban Development (SDAU) and Development Plans are both long-term documents (25 and 10 years respectively), while the majority of them were prepared without any accurate risk mapping (Table 4.3). Although they contain no-build zones, these are most often limited to river beds, in addition to an easement strip of 2 to 6 metres, which corresponds to the Public Water Domains established by the ABHs. They rarely include a map of flood expansion zones combined with regulated frequencies (e.g. a 100-year flood), or an indication of highest known water levels, such as in numerous OECD countries. Today, only the Al Hoceima SDAU contains an urbanisation suitability map, developed on the basis of the level of risk.

Granting building permits is the responsibility of municipalities. They must ensure laws and town planning regulations are enforced and stipulations in Development Plans and town planning documentation are respected, which makes them important actors in risk prevention. In addition to respecting town planning documents, the review of building permits also gathers various national public administrations, including Hydraulic Basin Agencies and Urban Agencies, whose notices are compliant prior to the municipality's decision. It is particularly at this level that the lack of precise and legally valid risk mapping becomes a roadblock for a proper implementation of non-structural preventive measures. The urbanisation suitability maps developed for the Al Hoceima Province, along with a dissemination brochure aimed at raising awareness of local stakeholders, represent a good example of "best practices" for other towns in Morocco to follow. In the absence of risk mapping, it is still possible to build in high-risk areas by taking advantage of legal loopholes. It is now envisaged to generalise the development of such maps of aptitude for urbanisation to the most vulnerable territories. Five pilot territories have already been identified within the Prefectures and Provinces of Driouch - Nador, Chefchaouen, Fnideq, Tangers - Fahs Anjra, and Taza.

Table 4.3. Overview of Master Plans for Urban Development in Morocco

Area affected	Year study launched	Date approved/ published in the *Bulletin Officiel*	Preparation time	Projected completion date	Update
Casablanca	Not provided (NP)	May/June 1985	NP	2005	Finalised
Rabat-Salé	NP	January/February 1995	NP	2010	Ongoing
Fez	NP	January/February 1996	NP	2010	Ongoing
Marrakesh	NP	January/February 1997	NP	2010	Ongoing
Agadir	1985	January/February 1998	13 years	2010	Ongoing
Tangier	1993	February/March 1998	5 years	2020	Ongoing
Settat	NP	January/February 1999	NP	2005	Ongoing
Chefchaouen	1993	February/March 1999	6 years	2020	NP
Tétouan	1993	March 1999	6 years	2020	Ongoing
Sidi Kacem	1993	September 1999	6 years	2020	NP
Al Hoceima	NP	January/February 2000	NP	2005	Ongoing
Tétouan Mediterranean coastline	NP	February/March 2000	NP	2020	NP
Nador	NP	January/February 2001	NP	2005	Ongoing
Tan-Tan/Tan-Tan Beach	1994	February 2001	7 years	2020	NP
Greater Ouarzazate	1994	May 2001	7 years	2020	-
Meknes	1992	September 2001	9 years	2020	Ongoing
Beni-Mellal and the irrigated sector of Tadla	1994	January 2001	8 years	2020	Ongoing
Taza		January/February 2002		2005	Ongoing
Khemisset-Tiflet	1994	October 2002	8 years	2015	NP
Essaouira	1993	October 2002	9 years	2015	Ongoing
Sidi Slimane, Sidi Yahya and Souk Larbaa (Grand Kenitra)	1995	October/Nov. 2003	8 years	2015	Ongoing
Tangier-Tétouan tourism coast	1995	March 2004	9 years	2020	Ongoing
Assilah and its coastal tourist strip	1994	May/June 2004	10 years	2020	NP
Tiznit and Sidi Ifni	1995	March 2004	11 years	2020	NP
Laayoune and Laayoune Beach	1995	January/March 2004	9 years	2015	NP
Northern Agadir coastline	1995	August/Sept. 2004	9 years		NP
Taroudant and Ouled Taima	1995	December 2004	9 years	2020	NP
Greater El Jadida	2000	June 2009	9 years	2020	NP
Ziz Valley	2002	October 2009	7 years	2025	NP
Daraa Valley	2002	September/Oct. 2010	8 years	2025	NP
Bouknadel-Moulay Bousselham coastline	2000	August/Sept. 2011	11 years	2020	NP
Central Mediterranean coastline	2000	March/April 2011	11 years	2020	NP
Todra and Dadès Valleys	2002	August 2011	9 years	2025	NP

Source: Ministry of Town and Country Planning

It should also be noted, the coverage of territories by urban planning documents in recent years has increased - with 24 SDAUs and 294 Development Plans developed between 2000 and 2015 - and a large number of the SDAUs are currently being updated. This offers excellent opportunities to better and more directly incorporate the issue of risk ahead of time in town planning documents. Putting together SDAUs is traditionally a long process. To guarantee the relevance of future town planning documents, it is important to speed up these updating processes in order to ensure the consideration of the economic, social and fast-changing urban context, as well as the most accurate risk assessment information. The development of new terms of reference that integrate more centrally the issue of risks in the new generation of urban planning documents and the publication of a guide on the 50 steps to be followed in developing these documents are also part of this approach to improve this essential aspect of risk prevention policies.

Construction codes and buildings' resistance to earthquakes

To cope with the risk of earthquakes, Morocco's paraseismic construction code has been in place for about a decade. The first paraseismic construction regulations in Morocco emerged in 1960, as a response to the devastating earthquake that struck Agadir. These were dubbed the "Agadir 1960 Standards". There was then a gap of 40 years before the national adoption of paraseismic standards in 2000, known as the "Paraseismic Construction Regulations RPS 2000". However, they did, not come into force until 2004, following the Al Hoceima earthquake. RPS 2000 defines "the rules for the calculation and sizing of structures to increase the stability of buildings during seismic tremors." It also dictates "the necessary technical provisions for civil engineering and architectural design in order to ensure buildings have optimum resistance to intense tremors" (Ministry of Housing and Town and Country Planning, 2011).

Moroccan paraseismic regulations have evolved over the years. In 2008, the Ministry of Housing launched a wide-ranging consultation process aimed at identifying some of the main lessons learned from the first years of the Regulations' application, which culminated in a revised version in 2011. A survey of professionals of the housing and construction sectors, including architects, engineering consultants, and real estate developers, helped shed light on some deficiencies in the way the Regulations were being enforced. This culminated in a revised version, and, in partnership with the Mohamed V University in Rabat, an update of seismic maps that supported the modelisation and parameters of the Regulations. The updated version of the Paraseismic Construction Regulations RPS 2000 took into account the increased number of building categories and types of geological sites, and amended the modelling of earthquakes' lateral force. Discussions with different partners also provided some insight on the additional costs of building standards, estimated by the Ministry of Housing at 10 to 15% of the structure's cost. Meetings held in the most affected regions helped raise awareness about these new codes among elected officials and representatives from the building trade. Experts and stakeholders both stressed the exemplary nature of such an approach.

Addressing existing issues

In light of the difficulties involved in the implementation of town planning policies, unregulated housing also tends to develop in high-risk areas, constituting a major vulnerability factor. People frequently build their own homes in Morocco, particularly those having migrated from rural areas. In 2010, 8.2% of urban households lived in shanty towns (UN-Habitat, 2012). In seismic zones, this makeshift housing is extremely vulnerable to risks, particularly to earthquakes. This form of housing is also often located in flood zones, such as the outskirts of Casablanca. Remarkable efforts have been deployed to clear these slums, along with measures to provide alternative housing. This was, for example, the case for areas surrounding Oued Bouskoura in Casablanca (see Box 4.7), or areas at the foot of mountain ranges in Agadir. Elsewhere, initiatives providing basic services (water, electricity) to these spontaneous settlements and other local protection measures have guaranteed a level of resistance for these communities. However, questions still arise around the long-term vulnerability of these neighbourhoods when confronted with more extreme events than those envisaged by protection-related projects, and the provision of basic services could end up tempting people to settle in high-risk areas.

The Paraseismic Construction Regulations "RPS 2000" have significantly contributed to strengthening the resilience of buildings in Morocco. However, they only apply to construction completed after they came into effect. It remains vital, therefore, to highlight the need for increased resilience of buildings constructed prior to the Regulations coming in; in particular, numerous ancient city centres, including the historic medinas, which are a large part of Morocco's cultural heritage and major tourist attractions. A number of public establishments like schools and hospitals may lack the required resilience to withstand an earthquake. It appears no steps have been taken to reduce exposure to the risk of tsunamis. Some people interviewed by experts went as far as saying Morocco must "turn much more towards the ocean than it did previously".

Box 4.7. Rehousing of communities vulnerable to flooding in Casablanca and Agadir

Recently, rehousing programmes for vulnerable communities living in flood zones have been put in place in Casablanca and Agadir by Moroccan authorities.

In Casablanca, the National Human Development Initiative (INDH) developed a specific shanty town eradication programme, managed by the real estate company Idmaj Sakane. Idmaj Sakane invested 1.6 million Dirhams to help rehouse 40,000 households, including those affected by the flooding in the Bouskoura slum in 2010. It built a broad network of players via a partnership-based approach, which has been able to obtain significant funding for the demolition of the most vulnerable shanty towns by ensuring all stakeholders work together. In this way, by rehousing communities that had settled in an unauthorised manner in specific areas, Idmaj Sakane has cleared land and created opportunities for real estate developers, which are consequently motivated to finance rehousing projects.

In Agadir, a foothills restructuring project attracted nearly 200 million Dirhams' worth of investment. It consisted of demolishing the existing shanty towns and rehousing over 400 households living in near-ruined homes in flood zones or zones threatened by the risks of landslide and subsidence. The upgrading of roads, the installation of street lighting, the provision of access to drinking water and the installation of a sanitary system are all considered to be ways of bringing a definitive end to the development of such shanty towns.

Sources: Interviews conducted as part of the OECD Study on Risk Management in Morocco (May 2015); Ministry of Economy and Finance (2014), draft budget bill for the 2015 financial year: Report on the special Treasury accounts, Ministry of Economy and Finance, www.finances.gov.ma/Docs/2014/DB/cst_fr.pdf; La Vie Eco (2014), Cities without Slums: final stretch for Casablanca, 7 May 2014, La Vie Eco, http://lavieeco.com/news/economie/villes-sans-bidonvilles-derniere-ligne-droite-pour-casablanca-29554.html#Tq1yK8XuCYagYj1z.99, consulted on 9 December 2015; L'économiste (2012), Agadir: Preparations Begin for the Foothills Restructuring Project, 20 February 2012, L'économiste, http://www.leconomiste.com/article/891419-agadir-la-restructuration-des-pi-monts-se-pr-pare, consulted on 11 February 2015.

Importance of control procedures

The presence of inhabited areas in flood zones is demonstrated almost every time flooding occurs in Morocco, raising the question of whether risks have been effectively taken into account in local development. In addition to regulatory loopholes and informal settlements, a number of building permits appear to be granted for high-risk areas. While strengthening controls and sanctions was part of the new law on urban development drafted in 2004, after the Al Hoceima earthquake, elected representatives reveal a strong preference for abandoning these measures. The clarification of responsibilities between local elected representatives and central administration requires adjusting the regulatory framework. However, this will not increase the extent to which risks are taken into account at local level without greater awareness of the risks and availability of clear information. Only then, can there be an effective control framework that requires stakeholders to face their responsibilities. Concerning informal and substandard housing, satellite mapping is a useful tool for town planners who can use the service contract proposed by the Royal Centre for Remote Sensing (CRTS) (see Box 4.8).

The effective implementation of the paraseismic regulations also requires robust controls to verify compliance. Generally speaking, upon the completion of a construction, an approval certificate is established by a building control office allowing municipalities to issue a permit to inhabit the premises. Building control offices are well-trained on these issues in the regions of Agadir and Al Hoceima, but less so for other regions equally exposed to the risk of earthquakes. The question of legal responsibility becomes an issue that causes legal difficulties in the event of non-compliance with regulations; indeed, upon completion of the construction, developers are generally no longer responsible for the land or buildings after these have been sold to a third party. More robust building control procedures throughout construction would help ensure construction stakeholders are held responsible. Putting compulsory signs on the façades of buildings that display the names of architects and developers is another way of ensuring greater responsibility.

Box 4.8. Use of satellite imagery to control urban sprawl

Driven by a massive rural exodus and by poverty, informal housing increases Morocco's vulnerability to natural disasters. To stop it from expanding, Moroccan authorities can use the services of the Royal Centre for Remote Sensing (CRTS) and satellite imagery to monitor its development, identify areas of possible expansion that are not permitted under the current regulations, and contain the sprawl in a relatively short period of time. The CRTS provides regular satellite images of key peripheral areas to the authorities (including the Ministry of the Interior and some local authorities).

For example, the *préfecture* of Inezgane-Ait Melloul concluded an agreement worth 1.41 million Dirhams with the CRTS, covering the 2010 to 2012 period, to obtain satellite images and maps showing the areas of urban expansion and illegal construction. Nevertheless, the effectiveness of this practice has been limited: in its 2013 annual report, the Moroccan Audit Office noted "maps and images provided by the Royal Centre for Remote Sensing to the *préfecture* of d'Inezgane-Ait Melloul have not been used". While the initiative is certainly interesting, building capacity of the agencies to interprete satellite imagery is required for a more effective contribution to the decision-making process.

Source: Interviews conducted as part of the OECD Study on Risk Management in Morocco (May 2015): Moroccan Audit Office (2013), Audit Office Annual Report (2013), Audit Office, http://www.courdescomptes.ma/upload/MoDUle_20/File_20_166.pdf.

Rapid urbanisation also raises the issue of increased soil sealing and the sizing of urban drainage systems. The systems have repeatedly shown their incapacity to evacuate storm water adequately after a new construction, as occurred in Tangier in 2008. This also raises questions about the standards adopted and the studies completed, as well as about the building work compliance controls. Environmental impact studies based on the application of Law 12-03 are also of critical importance, particularly to ensure new constructions do not increase the risk of flooding. Monitoring the implementation of the sizing recommendations contained in these studies is therefore essential to reduce the risk of flooding in urban environments.

Moving towards resilient cities

The ongoing decentralisation process offers new opportunities to improve the integration of risk management through town planning documents. Since 2003, and according to Organic Law 113-114 on municipalities, municipalities have been required to prepare a Community Action Plan (PAC). Within this framework, municipalities with extensive powers for risk prevention are required to present the planned actions at community level for the upcoming six years, based on a participatory initiative that supports sustainable development. The PACs, therefore, offer the opportunity to improve the participation of local actors in town planning mechanisms. The city of Tétouan, for example, has incorporated the issue of risk prevention at the heart of its PAC. The extension of similar participative approaches could be encouraged to better integrate resilience-related issues within urban development.

The implementation of ambitious "resilient cities" initiatives is an interesting approach to develop models that could serve as examples for various cities across Morocco. Through its 100 Resilient Cities initiative, the Rockefeller Foundation promotes a similar model and provides support to 100 cities in their attempts to become more resilient. (Box 4.9). A positive "resilient cities" approach based on innovation can harness synergies, since the issues relating to the sustainability of cities, the growth of green areas, and intelligent cities all lie at the heart of current town planning issues. The launch of the National Program for Sustainable Urban Planning by the Ministry of Town and Country Planning is part of this approach. Intelligent city approaches have been developed in Casablanca by LYDEC in particular, and economic incentives are allowing urban development to integrate questions posed by sustainable development. This includes the use of alternative techniques for slowing rainwater drainage in urban areas (valleys, eco-roofs) or making use of storm drain basins for the sustainable development of the city. More than 150 hectares have been treated in this way in Casablanca using alternative techniques that are often less expensive than reservoirs. The construction of six new towns by the Al-Omrane Group also represents an opportunity to integrate the question of resilience into a project from early design stages, as well as to test innovative pilot approaches that could serve as models nationwide.

Box 4.9. Rockefeller Foundation 100 Resilient Cities Initiative

Launched by the Rockefeller Foundation in 2013, the network of "100 Resilient Cities" (100RC) support cities across the world to adopt and implement a resilience plan within the framework of policies targeting global development and enhancement of territories. The objective is to enable members of the network to be better prepared for potential shocks, be they environmental, social or economic. Through the 100RC initiative, cities will receive financial and logistical support to create the new role of "Director of Resilience". This person will lead efforts to improve resilience and provide technical support for the development of a holistic resilience strategy that reflects the needs of each city. The network also provides access to a group of private sector and civil society partners that allows cities to develop and implement their resilience strategies. Finally, 100RC enables cities to take part in a forum where experiences and best practice can be shared.

Source: 100 Resilient Cities (2016), 100 Resilient Cities project website, www.100resilientcities.org/about-us#/-_/, consulted on 18 February 2015.

Other non-structural prevention measures

Drought prevention in agricultural areas

The farming sector – still of major socio-economic importance in Morocco – is particularly vulnerable to drought. In such a context, the *Plan Maroc Vert* (Green Morocco Plan) aims to reduce the reliance of the country's agricultural sector on climatic fluctuations. It also includes regeneration projects to replace land use types involving high water consumption with high added-value crops that are less sensitive to the lack of rainfall. Within this regeneration programme, an average of 12 million fruit trees have been planted since 2009. In addition, major investment in irrigation – part of the water saving programme – is underway (37bn Dirhams invested between 2008 and 2020), and assistance is being provided to farmers to move towards crop management techniques that consume less water and improve their competitiveness. Finally, incentives encouraging farmers to adopt good farming practices and appropriate crop management methods are being implemented through the Agricultural Development Fund.

Assessment of ecosystems for risk management

To reduce the risks of flooding and drought, Morocco has opted for flexible measures based on value assessment and strengthening of services provided by ecosystems. Programmes are conducted within the framework of the fight against desertification or the conservation of water and soil. These programmes include the protection of river basins, reforestation, and measures to combat erosion or overgrazing. They help reduce the risks of drought, flooding and landslides by retaining more rainwater in areas upstream from the basins. The contribution of work in these areas by the High Commission for Waterways and Forests and the Fight against Desertification, the Ministry of Agriculture and the Hydraulic Basin Agencies would benefit from greater visibility at the heart of a national strategy for risk management.

Resilience of crucial sectors and businesses

Ensuring the resilience of critical infrastructure sectors is a major challenge for an integrated risk management policy. This is particularly important in a context of significant infrastrucrture development in Morocco. The impact of natural disasters on critical infrastructure (water, telecommunications, energy and transport) can have knock-on effects for a wider area than the one directly affected by the disaster, and for a much longer period. This leads to more significant damage, which can affect the well-being of communities and the economy. The economic impact can also be exacerbated by the interruption of production processes of companies affected by the disaster. This can also have adverse effects on their relationships with clients and suppliers across the entire value chains. The increased vulnerability of modern companies to this kind of damage is at the heart of the OECD's work on risk management and its Recommendation on the Governance of Critical Risks.

Critical infrastructure and resilience in Morocco

Recurrent flooding in Morocco in recent years has shown the significant impact of disasters on the country's critical infrastructure. In November 2014, almost the entire province of Guelmim was cut off from the national road network, making it more difficult for assistance to arrive and forcing the authorities to organise supplies via sea routes. Flooding in the town of Agadir swept away one of the two water pipes managed by the National Water Authority and it could have threatened the city's drinking water supply. In 2010, the railway line connecting the financial capital of Casablanca and the political capital of Rabat was badly affected, leading to significant economic losses. The risk of drought can also have significant consequences for water supplies, as demonstrated by the emergency supply provided to Tangier by tankers during the drought of 1995.

The Ministry of Public Works, Transportation and Logistics (METL) is particularly involved in current risk management efforts in Morocco. It has a risk management and prevention plan of action that is specifically dedicated to the transport sector. It uses the latest techniques to design infrastructure, principally basing these on 100-year flood levels and in close adherence to paraseismic regulations. The Ministry has also launched an initiative with operators responsible for roads, ports, airports and railways to reflect on the organisation and continuity of services and on the consolidation of the Ministry's capabilities and its agencies. This has led to a risk assessment of infrastructure, including the impact of climate change on certain pilot projects. Major infrastructure development projects also present opportunities to improve Morocco's approaches to the resilience of critical infrastructure; For example, the high-speed line enables the National Railways Authority (ONCF) to incorporate the best international standards in terms of seismic studies and the hydraulic transparency of this infrastructure. Crisis management skills are also constantly improving (Chapter 5). Nevertheless, these initiatives have not yet led to the development of an investment plan dedicated to risk prevention and to improving the resilience of existing infrastructure. No study has been specifically conducted on the risk of tsunamis despite the location of significant public works in coastal regions.

Other sectors of critical infrastructure are generally managed by the private sector. This is particularly true for the water and electricity distribution sector and for telecommunications (Table 4.4). While the law on the delegated management of public services contains elements about the requirement to manage companies operating in the water and electricity sectors to ensure service continuity, there are no regulations containing detailed provisions for these sectors of critical infrastructure. In the telecommunications sector, the Telecommunications Regulatory Agency's specifications reiterate the continuity of service requirement but they do not provide further details. In such a context, it is often up to operators to adopt voluntary resilience-related approaches to meet these obligations. Several examples show how they integrate service continuity into their strategies. In the case of Maroc Telecom, this is achieved through the redundancy of networks, while public and private operators in the water distribution sector use an integrated approach that combines investment in prevention with risk management methods.

Table 4.4. Principal operators of critical infrastructure in Morocco

Sector	Key player	Role	Status	
			Public	Private
Water	National Office for Drinking Water	Production/Distribution	¤	
	Hydraulic Basins Agencies	Dams	¤	
	Distribution management	Distribution	¤	
	LYDEC	Distribution		¤
	REDAL	Distribution		¤
	AMENDIS	Distribution		¤
Transport	National Airports Authority		¤	
	Motorways of Morocco		¤	
	National Port Authority		¤	
	Tangier Med Port Authority			¤
	National Railways Authority		¤	
Energy	National Electricity Authority	Production and Distribution	¤	
	LYDEC	Distribution		¤
Telecoms	Maroc Telecom	Operator		¤
	Médi Telecom	Operator		¤
	Wana Corporate	Operator		¤
	National Telecoms Regulatory Authority	Regulation	¤	
Banking System	Bank Al Maghrib	Issuing of money, Regulation	¤	

Source: Interviews conducted as part of the OECD Study on Risk Management in Morocco (May 2015).

These voluntary approaches, or those resulting from the desire of operators to align with international standards, are not supported by a specific regulatory framework or initiative. The experience of OECD countries shows the value of creating a specific framework for operators of critical infrastructure given their diverse nature and varied status. The development of such a framework would make it possible to ensure various sectors of critical infrastructure follow a standardised approach at national level and guarantee shared objective of resilience. It is worth remembering that these operators may occasionally be in competition with one another, or operate in different sectors, but they are often interdependent. This would help improve the resilience of critical infrastructure, an issue of growing importance today. The creation of an appropriate governance framework in which regulators and operators can discuss the service continuity levels to be guaranteed is an approach generally adopted in numerous OECD countries. A variety of initiatives or regulatory approaches can then be applied, as appropriate (Box 4.10).

Box 4.10. Examples of policies to promote the resilience of energy infrastructure

In the United Kingdom, the Energy Law of 2013 allows the Department of Energy and Climate Change to charge in return for energy resilience services in the event of an interruption to the power supply, whether planned or the result of a fault.

In Finland, the National Authority for Emergency Supply promotes the continuity of activity and resilience through public-private partnerships. A dedicated energy sector department offers specific tools like IT systems or storage and transport infrastructure to support the continuity of business activity. This agency also develops incentives to protect critical infrastructure and prepares annual reports on the performance of energy firms, including a rankings table and specific recommendations.

In the United States, the Federal Hurricane Sandy Rebuilding Task Force published a "Strategy for rebuilding after Hurricane Sandy" in August 2013 that included precise recommendations to guarantee that investments for the recovery of the energy sector were resilient (Recommendation 12) and that they would encourage co-operation between central and federal authorities to improve the policies and standards in the electricity network (Recommendation 14).

Source: OECD (2015) Establishing effective Public Private Partnerships for risk management, What are the possible options for government?, internal working document, Directorate for Public Governance and Territorial Development, OECD, Paris.

Resilience of businesses

The question of resilience to natural risks is a new challenge for businesses. Significant economic damage was caused to businesses by major flooding in Casablanca in 2009, or as a result of poor installations in industrial districts. In 2008, 200 businesses suffered damage and significant losses of production, in addition to the direct damage, caused by the disaster in Tangier's industrial district.

Relocating industrial activities situated in flood zones or in areas at high risk of earthquakes is a rather sensitive issue, as it is extremely costly for businesses. If public authorities wish to contribute to financing such actions, it becomes equally onerous on the country's public finances. In such a context, incentives to relocate can nevertheless be offered in the form of tax credits. For insured businesses, incentives can also take the form of a reduction in insurance premiums if they choose to relocate. In such a case, insurance companies must take on an informative role (Box 4.11).

When this is not feasible, other measures can be adopted, particularly by SMEs. The window of opportunity available after a disaster must be used to encourage more voluntary prevention measures. A self-diagnosis, either with or without the help of consultancy services, can be carried out. For example, some questionnaires attached to case studies can be used as a self-diagnosis tool, and relatively low-cost measures can be adopted by manufacturers. The same is true with the risk of earthquakes; a company can ensure that hard pipes for water and steam, and pipes supplying materials and chemicals are replaced by pipes made from flexible material in all partition wall passageways. For the risk of flooding, the same type of specific measures can be considered: setting up the electricity network command on the first floor (umbrella network), and ensuring engines and machinery can be rapidly dismantled and installed on the first floor. Transport companies should be able to move their fleets of lorries to an area away from the flood-risk zone. All businesses should begin to conduct regular exercises.

Box 4.11. Diagnosis of the vulnerability of businesses in the Loire Basin

In the Loire Basin, a flood-risk prevention action programme aimed at businesses conducted by the Loire River Basin Authority (EP Loire) recently contributed to informing more than 15,000 companies, despite a difficult financial and economic environment. Based mainly on a free vulnerability diagnosis for businesses located in flood-risk areas, the aim of this "industrial" initiative was to significantly reduce the vulnerability of economic activity in the Loire Basin and its tributaries by preserving the vital interests of businesses in the event of major flooding.

Meeting this objective provided businesses with a useful diagnostic tool in the decision-making process and enabled awareness raising of the risk of flooding among more than 15,000 companies. Two thousand vulnerability diagnoses were requested by companies, 1,812 of which were already completed or in progress in 2012. The diagnosis included an on-site analysis and the delivery of a report identifying vulnerable areas. These were listed in order of seriousness with an estimate of the costs of the potential damage. Businesses that were interested in receiving a vulnerability diagnosis were then offered assistance to assess the need for effective measures that reduce their vulnerability, along with an estimation of the cost that such measures would involve. In some cases, the identified measures could benefit from co-financing. In 2011, 280 companies were monitored in this way, 21 of which received financial support.

Source: OECD (2014) *OECD Study on the Management of Flood Risks: the Seine in Paris Region 2014*, OECD Publishing, Paris.

Under the influence of their leadership, some large private companies took the initiative to embark on significant prevention initiatives. International regulations and standards linked to risk management (ISO 31000) and business continuity (ISO 22301) and the raising of awareness among suppliers and international clients or insurers have all contributed to the engagement of businesses with an international dimension. Beyond infrastructure, some sectors have been able to structure their efforts around a business leader. For example, the banking sector has committed to an ambitious resilience initiative under the leadership of the Moroccan Central Bank (Box 4.12). Large Moroccan groups are gradually embracing these subjects, such as the Office Chérifien des Phosphates (phosphate rock producer) that invests in Casablanca's resilience by financing the Super Collector West project (Box 4.5).

The private sector is not supported by a framework that incentivises the improvement of prevention and it suffers from a lack of risk-related communication at national level. This is particularly the case for small and medium-sized companies. Despite the fact that continuous access to logistical corridors is a significant competitive asset, the development of several industrial districts, particularly in the framework of an Emergency Industrial Plan, is not supported by a risk analysis.

Box 4.12. Morocco's Central Bank supports business continuity in the banking sector

Bank-Al-Maghrib, Morocco's central bank, supports business continuity of lending institutions. Since 2008 in particular, the Directive relating to the business continuity plan within lending institutions has specified the minimum rules that must be observed by such establishments in the implementation of a business continuity plan. Bank-Al-Maghrib's initiative is part of the implementation framework of the Basel II Accords and it is inspired by international best practices in risk management (ISO 31000) and business continuity (ISO 22301).

The regulations published relate in particular to the policies and responsibilities regarding business continuity and methods for implementing crisis scenarios and impact analyses. The Directive specifies the components to include in the business continuity plan, such as strategies and procedures for protecting and recovering data, procedures for the emergency recovery of data, applications and important material, or the identification of alternative replacement sites. The business continuity plan must also include protocols for internal and external emergency communication, as well as a reporting mechanism to the Bank-Al-Maghrib.

The Bank-Al-Maghrib is implementing its own parallel Business Activity Continuity Plan (PCA) in order to ensure the continuity of its most critical activities and minimise the impact of major operational disturbances on people, assets, the financial system and the reputation of the Bank. This is highly important since the Bank-Al-Maghrib's activities were hampered by recent natural disasters that hit the country. During the Guelmim flood of 2014, the bank was forced to take preventive measures when faced with the risk of an interruption to cash supply transactions in the affected areas.

The PCA of the Bank-Al-Maghrib contains a decentralised model of governance dedicated to crisis situations, continuity plans for all bodies, an IT continuity plan, a crisis management plan, and a maintenance plan for an operational state, with a programme of tests and exercises. It aims to protect the Bank-Al-Maghrib against threats from natural causes (floods, fires, earthquakes, tsunamis, epidemics and pandemics) but also against technical causes (power cuts, breakdown of IT systems) and human causes (vandalism, human errors, and attacks on IT systems). It covers a variety of scenarios, including a major incident affecting an area where the Bank operates or an incident directly affecting the premises of the bank, and it organises back-up exercises, crisis simulations and emergency IT recovery tests in order to guarantee its operational status.

Source: Bank Al Maghrib (no date), directive relating to the business continuity plan of activity within lending institutions, www.bkam.ma/wps/portal/net/kcxml/04_Sj9SPykssy0xPLMnMz0vM0Y_QjzKL94w3cQwBSZnFe8QbebvqR2KIuSDEfD3yc1P1g9KK9L31A_QLckMjyh0dFQHflhUF/delta/base64xml/L0lDU0lKQ1RPN29na21DU1Evb0tvUUFBSVFnakZJQUFRaENFSVFqR0VKemdBIS80SkZpQ28wZWgxaWNvblFWR2hkLXNJYnpRQSEhLzdfSV84TUgvOQ!!?cont574908=43&formid=, consulted on 14 December 2015.

The relocation of existing industrial districts is a possible option if corrective measures are not feasible. In this case, areas with categorised industrial establishments should be over-sized to allow sufficiently safe distances between establishments and ensure alleyways are wide enough for evacuation, even if this affects the cost price. Plans should also include shared fire-fighting services and direct telephone lines to civil protection agencies. The new regulations will require danger and impact studies be carried out before granting a licence to operate. Investors must take a long-term view of any additional costs (improved resilience, better company image, reduced upgrade costs at a later stage).

Natural risks can result in damages for critical infrastructures. However, critical infrastructure can also cause accidents of equal or greater significance. It is, therefore, crucial Morocco rapidly equips itself with a modern regulatory framework. As noted in Chapter 1, the law of 1914 needs to be reviewed and enhanced to deal with the complex and evolving industrial risks. A useful example is the French Law of 2003 on the prevention of technological and natural risks (Box 4.13). Like in Morocco, progress on this front was driven by disasters; this law was created as a result of the serious AZF industrial explosion in Toulouse in September 2001. The law is mostly relevant for organisations covered by the European Seveso Directive (high threshold).

Box 4.13. French Law No. 2003-699 of 30 July 2003 on the prevention of technological and natural risks and the repair of damage

Also known as the Bachelot Law – named after the erstwhile Minister of the Environment – this law is extremely ambitious and wide-ranging. It amends several areas of legislation (environment, town planning, civil, commercial, employment, insurance, rural affairs, road safety, forestry and taxation laws), given that natural and technological risks cut across a wide field. The law establishes plans for the prevention of technological risks (PPRT) which contribute to the management of urban areas surrounding establishments at risk.

The operators (both existing operators and applicants) must provide public authorities with a "study of the dangers specifying the risks to which the installation may be exposed, directly or indirectly, including those relating to hygiene, public health and safety in the event of an accident caused either inside or outside the installation" (Article 4). The study considers the likelihood of such an event occurring, the kinetics involved and the seriousness of potential accidents. The study also includes an estimate of the potential cost of material damage to third parties in the event of an accident. The operator must send this assessment report to the local *préfet* and local risk information and consultation committee (Article 21).

A PPRT can be prepared on the basis of these studies. It outlines the scope of the study for which the PPRT can provide various measures, including those relating to property (expropriations, right of abandonment, etc.), additional measures to reduce risk at source (changes in procedure, etc.), reinforcement work on neighbouring buildings, and restrictions on future urban development. This law is an original piece of legislation because it specifies the role of the various stakeholders as well the means for financing the PPRT: either a friendly agreement is established between the operators, the state and regional authorities, or if this is not possible, a sharing of tasks is established by a third party.

Source: Taken from "French Law No 2003-699 of 30 July 2003 on the prevention of technological and natural risks and the repair of damage".

The law of 30 July 2003 is also a reference text for floods; although it envisages very high insurance premiums (50% or more of the cost of buildings), the state and local authorities can supplement the insurance payment to allow the relocation of buildings, rather than repaired and left on a flood-risk site. Based on these principles, and on the right of expropriation, homeowners at Faute-sur-Mer in France were able to receive compensation after their homes were flooded during storm Xynthia in February 2010.

Following a decade of implementing this law, a report from the French Ministry of the Environment, Energy and Maritime Affairs highlights lessons that could be useful for Morocco. It took ten years to build ownership, implement site inspections, and achieve harmonisation. As a result, methodology guidelines could be developed. Plans for 407 industrial districts involving 800 communities have specified the responsibilities of various stakeholders. Property measures have affected 10,000 people and involved 2bn Euros of investments. Industrial operators have invested 200 to 300 million Euros annually. The reverse side of the coin: financing by the state and local authorities was considerable and directly proportional to the law's success. As a result, Law N°2013-619 was required to introduce limits to construction work financing for existing buildings, while offering individuals tax credits and additional help.

The lack of information among companies on risks and on business continuity can hinder their awareness and limit their commitment. Organisation by the General Confederation of Moroccan Companies (CGEM) of conferences on risk management in Rabat shows a promising start. More significant formalisation of these initiatives could be inspired by the *Club des Directeurs de la Sécurité en Entreprises* in France (Box 4.14).

Box 4.14. *Club des Directeurs de la Sécurité en Entreprises*

Since the beginning of the 1990s, the *Club des Directeurs de la Sécurité en Entreprises* (CDSE) has gathered about a hundred of France's largest companies in the frame of an association.

CSDE's main objective is to provide member companies' security and safety departments with the tools to study, communicate, intermediate, and train, as well as the expertise required for their missions on issues relating to the protection of people, company materials and non-material assets, business intelligence and more generally, crisis anticipation and management.

The CDSE is supported by tools such as thematic working groups, and the publication of a high-level professional magazine and a periodical for safety managers. It certifies and organises training sessions, publishes a safety barometer of international associates, and interacts with public authorities.

As a result of the ever-increasing responsibilities of its members (data processing, product traceability, risks with regards to beneficiaries), and the privatisation of a number of activities in sectors of vital importance, the responsibility shared with public authorities has led to an increasing number of important partnerships with the Ministry of the Interior, Justice and Defence, and Foreign Affairs, often through formal agreements. The CDSE regularly uses OECD tools as a source of inspiration and comparison.

Source: CDSE (2016), *Club des Directeurs de la Sécurité en Entreprises* website, https://www.cdse.fr, consulted on 16 February 2016.

Financing prevention

Investments in risk prevention policies are often more effective than financing recovery and reconstruction measures following a disaster (World Bank, 2010). However, financing for prevention policies should also be allocated as efficiently as possible so it is important to obtain a clear vision of the investments in risk prevention, even if such an exercise can present a number of difficulties.

Wide-ranging financing options for risk prevention in Morocco

In Morocco, expenditure for prevention is often incorporated in diverse programmes covering various areas of public policy and can be managed by numerous stakeholders in central government, regional authorities and the private sector. This can include the financing of protective infrastructure, the costs involved in their operation and maintenance, salaries of civil servants in town planning administrations, the budgets of scientific and technical institutions, and additional costs of construction relating to the implementation of the paraseismic code. Based on best practices from countries that report all spending on risk management, such as Australia, Japan or Switzerland, the OECD has started to develop a framework to analyse risk prevention-related expenditure in order to help countries develop efficient prevention policies (Appendix H and I).

Morocco, like many countries, doesn't have a tool for identifying expenditure allocated to risk prevention. Such a tool would ideally provide a solid basis for developing an integrated policy. In fact, a wide range of financing tools is used for various risk prevention actions by different government departments involved in the design and implementation of such public policies.

This chapter refers to several programmes developed in Morocco in this domain. Table 4.5 shows various programmes conducted at national level. Most importantly, it includes the Fund to Combat the Effects of Natural Disasters (FLCN), which is dedicated to managing risks, and is increasingly directed at funding prevention. Funds from the *Plan Maroc Vert* (Green Morocco Plan) must also be considered with regard to the risk of drought. Concerning flooding, infrastructure investment by the Ministry of Water through its large dam programmes or its National Flood Protection Plan are supplemented by the Ministry of the Interior's National Liquid Sanitation Programme, which focuses on urban drainage. The Ministry of Works' programme is another important source of financing for risk prevention. The "Cities without Slums" Programme and other instruments from the Ministry for Housing, or its agency known as Al-Omrane, have also contributed to relocating communities living in at-risk areas. It is important to consider all of these programmes when creating a useful tool in the decision-making process. Some of the programmes can mobilise significant funding: although the Fund to Combat the Effects of Natural Disasters remains the main tool for financing prevention, its current resources do not reflect this ambition, with an average annual financing of 200 million Dirhams. This is three times as high as the financing from the National Flood Protection Plan (PNPI), which reached 7.6bn Dirhams over a period of 13 years, not including dam projects.

Co-financing contributions from regional authorities, research and operating budgets of scientific and technical institutions, and private sector expenditure oriented towards building its own resilience – like the financing of more extensive prevention measures (Super Collector project) or meeting additional construction costs to comply with paraseismic regulations – also contribute to financing risk prevention. This dispersal of financing does not make the sums allocated or the resource prioiritisation mechanisms clear. Developing a dashboard for risk management in Morocco would be a useful tool to support the national risk management strategy currently underway. Such a framework could be inspired by Australia, Japan or Switzerland (Appendices H and I).

An innovative practice: the FLCN's call for projects

Beyond the mobilisation of resources, a good risk prevention financing policy must address their allocation. In the context of recent moves aimed at strengthening risk management in Morocco, an innovative device for financing prevention projects based on a call for projects open to the various Moroccan government departments was launched in 2015 within the framework of the FLCN. When it was created in 2009, the fund was initially for the essential financing of post-disaster reconstruction. Since then, it has gradually moved towards financing prevention.

Table 4.5. Estimates of primary means of financing for risk prevention in Morocco

Name	Managed by	Targeted Risks	Examples of Prevention Measures Financed	Programme Total	Amounts Allocated to Prevention
Green Morocco Plan	Ministry of Agriculture and Maritime Fishing	Drought	Support for grain production conversion	20 billion Dirhams	20 billion Dirhams
			National irrigation water conservation programme	37 billion Dirhams	37 billion Dirhams
		Climate Risk	Crop insurance	250 million Dirhams annually	250 million Dirhams annually
		Extreme temperatures and hail	Incentives from agricultural development funds	Not communicated (NC)	NC
PNPI	Ministry of Water	Floods	Flood protection projects	25 million Dirhams	7.6 million Dirhams
Fund to Combat the Effects of Natural Disasters	Ministry of the Interior	All	Flood protection projects in 6 high-risk provinces, Western Casablanca Super-Collector Project (SCO)	200 million Dirhams annually	NC
Liquid Sanitation and Wastewater Treatment Funds	Ministry of the Interior	Floods	Improvement of overall coordination, development of water removal systems	714 million Dirhams (2016)	NC
Housing and Urban Development Funds	Ministry of Housing and Urban Policy	Earthquakes, Floods	Removal of unsanitary housing, at-risk housing, preventive housing activities	2 billion Dirhams (2016)	146,165 Dirhams
Cities Without Slums Programme	Ministry of Housing and Urban Policy	All	Eradication of shanty towns	32 billion Dirhams	NC
Special funds for the promotion and support of civil defence	Ministry of the Interior	All	Prevention programmes	200 million Dirhams annually	75,558 Dirhams
Rural Development Funds	Ministry of Urban Affairs and Planning	Earthquake, Floods	Preparing urban planning documents and structural plans	1.3 billion Dirhams annually	NC
National Forestry Funds (FNF)	HCEFLCD	Management of catchment basins, anti-desertification activities	Prevention programmes	200 million Dirhams annually	NC

Sources: Ministry of Economy and Finance (2015), Draft Bill for Finances in the 2016 financial year: presentation note, Ministry of Economy and Finance http://www.finances.gov.ma/Docs/DB/2016/np_fr.pdf; Mourahib, B. (2008), "Cities without Slums" does not help women, Telquel, 28 January 2015, http://telquel.ma/2016/01/28/programme-villes-bidonvilles-profite-pas-femmes_1480003, consulted on 18 February 2015; L'économiste (2008), 20 billion Dirhams for the Green Morocco Plan, L'économiste, n°2894, 3 November 2008, http://www.leconomiste.com/article/20-milliards-de-dh-pour-le-plan-maroc-vert, consulted on 18 February 2015.

This first call for projects constitutes an interesting practice to favour the emergence of prevention project supporters (particularly at a local level) through the use of financial incentives. The mechanism offers to fund 30% of the total cost of the prevention projects presented by the project supporters, who must raise 20% of the funding. Co-financing can represent as much as 60% for communities with fewer than 35,000 inhabitants and limited resources or for projects aimed at reducing the risk of recurrent disasters. It is based on precise terms that outline the objectives, the eligibility, and the selection criteria. A solid governance mechanism composed of a management committee, a national selection committee, and a small and functional secretariat ensures the transparency in the funds' allocation. The call also prioritises projects that cut cross different areas and involve institutional co-operation.

The call for projects has generated strong interest among stakeholders. The Fund secretariat received 90 projects, 80 of which were put forward by regional authorities, six by ministerial departments, and four by public establishments. This demonstrates the hopes and strong need for financing among local stakeholders in the area of risk prevention. This call for projects, therefore, represents an important step towards promoting further approaches to risk management. In June 2015, 23 projects of the 90 received were pre-selected and submitted to the national selection committee for a decision. Seventeen of those came from regional authorities, three from ministerial departments, and three from public establishments. This involved a total of 1.038bn Dirhams in investments, 24% (about 251 million) of which were financed by the FLCN. Such a leverage effect indicates the strong relevance of this recently introduced tactic.

This first experience showed the dominance of structural measures compared to non-structural measures, which represented only 4% of the proposed projects. The need to enhance the skills of local project supporters in non-structural measures, and more widely in prevention, is another lesson learnt. More than 50% of the projects proposed were actually ineligible, and therefore eliminated. The call for projects has also allowed the secretariat to assemble elements relating to risks in Morocco. This is particularly important for the national risk management strategy mentioned in Chapter 2.

Ultimately, all projects that satisfied the eligibility and relevance criteria were accepted for financing. It has, therefore, not been necessary at this stage to select projects using prioritisation criteria. Other countries like Austria, France and Mexico have introduced similar procedures (Box 4.15) and have also launched mechanisms that are less selective in order to encourage the development of local project management. From a long-term and sustainability perspective, it is often necessary to introduce prioritisation criteria based on cost-benefit analyses and risk assessments at a national level to ensure the most efficient risk reduction projects are carried out. It will then become a matter of introducing arbitration procedures that account for efficiency and fairness to allocate financing, while simultaneously maintaining the attractiveness of this incentivising device.

Box 4.15. Risk Prevention Funds in Mexico, France and Austria

The Katfunds in Austria, the Barnier Fund (or Fund for the Prevention of Critical Natural Risks) in France, and the FOPREDEN Fund (*Fund for the Prevention of Natural Disasters*) in Mexico, are all specifically designed for the prevention of the risk of natural disasters.

As is the case in Morocco, these funds stemmed from funds that initially focused on financing emergency responses. Since their creation, these funds' resources have gradually extended to cover measures aimed at reducing the vulnerability to risks. They have developed more precise selection processes for risk prevention projects, and they are therefore at the stage of developing processes to prioritise the allocation of funds. This is a classic approach that incentivises local stakeholders to gradually develop local project management skills in the domain of risk prevention, before selecting projects and orienting investments where they can be most effective.

Source: OECD (2013), *OECD Reviews of Risk Management Policies: Mexico 2013: Review of the Mexican National Civil Protection System*, OECD Publishing, Paris; OECD (2014), *Boosting Resilience through Innovative Risk Governance*, OECD Publishing, Paris.

Conclusion and recommendations

Ultimately, it seems that risk prevention is a public policy that needs to be developed and consolidated in Morocco, particularly by making local actors increasingly responsible. Strengthening prevention through co-construction and responsibility sharing will assist in moving beyond structural measures, which are costly and are now reaching their limits.

Non-structural measuresshared by local stakeholders constitute an essential condition to ensure prevention efforts are successfully implemented and can move beyond the current top-down approach. If this is accompanied by practical measures (e.g. community action plans, calls for projects), such an approach should be formalised, along with an enhancement of incentive and control measures to ensure the effective implementation of local prevention policies.

Strengthening the risk culture through awareness campaigns targeting communities, businesses and local actors is also a key element to encourage this approach. These awareness programmes could be incorporated into a national campaign for training public officials in order to ensure risk policies are disseminated and consistently understood across the country. The idea of using the school system, media and associations as vehicles for training and information could also be pursued.

Local experts and peers interviewed for this study noted the risk of tsunamis does not seem to be the subject of sufficiently specific risk prevention measures, despite the fact such an event could have extremely significant consequences for Morocco and could have long-term effects on the economy and society. A discussion of this risk raises the question of the proper place of the ocean and the sea in Morocco's future development.

The diversity of current prevention financing sources in Morocco seems to be a factor that could lead to a short-term extension of means. Strengthening non-structural prevention measures also depends on priority-based arbitrations. In the longer term, however, balancing means with prevention needs could prove difficult.

Recommendations

ESTABLISH A PERMANENT PREVENTION POLICY TO IMPROVE THE CURRENT STATE OF AFFAIRS AND STRENGTHEN LOCAL PROGRAMMES

Preserve and reinforce prevention financing via the FLCN, and prioritise funding and activities.

Developing a dashboard that compiles all risk prevention expenditure will help ensure the best prioritisation for funding based on risks and regions, as well as a more detailed assessment of each agency's contributions and their effectiveness. The incentive approach of FLCN's call for projects should be reinforced with an ongoing financing mechanism and progressive integration of selection criteria based on cost-benefit analyses. The development of a methodology and guidelines for project initiators will contribute to facilitate this type of arbitration.

Improve the way structural and non-structural activities are balanced in order to optimise investment and the use of public funding.

An update of the Ministry of Water's National Flood Protection Programme must highlight the evolution of priorities for flood risk and integrate the perspective of climate change. It is important to better combine structural approaches – which have been the priority until now – with non-structural approaches related to specific vulnerabilities that have come to light and to catchment basins. Good infrastructure governance involves the integration of cost-benefit analyses into decision-making criteria. This favours approaches that astutely combine multiple usages and takes better account of associated running and maintenance costs.

Initiate a national risk communication campaign focused on local responsibility.

A communication strategy on multidirectional risks should be built on the basis of an improved understanding of risks in Morocco. It should target local decision-makers and provide them with specific information about the risks their regions are subject to and measures to be taken to inform local residents.

This strategy also encourages improved awareness by businesses, specifically small businesses, about the issues involved in the continuity of operations. There are several possible approaches to this strategy: sharing experiences, information sharing within clubs, self-assessment via guidelines, the promotion of "contingency plan" insurance policies, financial incentives from the public sector, or just simple preventive measures (elevating the electrical system, enabling water to flow through pipes made from flexible material in all partition wall passageways in factories, removing motors, machines and computer systems, drills, adding a tsunami escape hatch in the roof, etc.).

Enhance the integration of resilience into regional and urban development through transparent and binding zoning policies.

The creation of urbanisation capacity maps, and their inclusion in urban and regional planning documents, will highlight the issue of risk in community action plans in order to combine bottom-up and top-down approaches. The development of resilience models in new towns or through the selection of a Moroccan city for the Rockefeller Foundation's 100 Resilient Cities programme will enable innovative resilience solutions to be adapted to the Moroccan environment and to be tested before they can serve as examples nationwide. Strengthening regional authorities' risk prevention capabilities will be a cornerstone of this strategy. Empowering and training the key players in the construction sector (architects, engineers, town planners and developers) in urban resilience approaches should also be considered.

Establishing a partnership with crucial infrastructure operators to include them in managing critical risk.

Strengthening the resilience of crucial infrastructure when faced with critical risks will require the development of a dedicated partnership between operators and government agencies. The objective is to ensure a transparent exchange of information on vulnerabilities and the desired levels of resilience. Public-private partnership agreements or regulations of critical infrastructure sectors must include specific clauses for the continuity of operations in critical or vitally important industries.

Reflect more actively about the prevention aspects of the risk of tsunamis.

Faced with a low-probability risk that can result in very significant damage to the country, a national programme for tsunami risk prevention should be initiated on the basis of a deeper reflection about Morocco's coastal development. The law on coastlines currently under discussion represents an opportunity for such a debate, which should focus on the long-term diversification of development strategies.

Bibliography

Bouregreg and Chaouia Hydraulic Basin Agency, "Executive Summary", *Draft master plan for the Integrated Management of Water Resources of the Bouregreg and Chaouia Basin*, NOVEC, http://www.abhatoo.net.ma/content/download/23190/428545/version/1/file/plan_directeur_amenagement_integre_ressources_eau_bouregreg_chaouia.pdf.

Oum Er-Rbia Hydraulic Basin Agency (2012), "Summary", Draft master plan for the Integrated Management of Water Resources of the Oum Er-Rbia Basin and Atlantic Coastal Basins, Oum Er-bia Hydraulic Basin Agency, http://www.abhoer.ma/pages_externes/PDF%20pdaire%20francais/R%C3%A9sum%C3%A9%20d%C3%A9finitif.pdf.

Casablanca Urban Agency (2008), Greater Casablanca Master Plan for Urban Development, http://www.auc.ma/docs/124112008133350.pdf.

Bouregreg Valley Planning Agency, draft legislation by the Bouregreg Valley Development Agency, www.bouregreg.com/tiki-download_file.php?fileId=15.

Architecture du Maroc (2013), "The SDAU, a new vision for the development of Casablanca", Architecture du Maroc, n°53, May-June 2013, pp. 28-34.

The Souss Natural Sciences Association (ASVTS) (2014), "2013/2014 والأرض بسوس لموسم اللقاء التواصلي لجمعية علوم الحياة" (Meeting of the Souss Natural Sciences Association for the 2013/2014 season), presentation during interviews conducted as part of the OECD Study on Risk Management in Morocco, June 2015.

Souss Natural Sciences Association (ASVTS) (2015), "Project: Princess Lala Aicha 'AGLA' educational seismic station", presentation during interviews conducted as part of the OECD Study on Risk Management in Morocco, June 2015.

Casa Aménagement (2014), Super Collector West website, http://www.casa-amenagement.ma/fr/nos-projets/super-collecteur-ouest, consulted on 21 December 2015.

Urban Municipality of Tétouan et al. (2010), CDP Tétouan 2011-2016 – Communal Development Plan of the Urban Municipality of Tétouan, http://www.medcities.org/documents/10180/210014/plan+communal+tetuan/3c4a6ce3-e2b4-4b68-b993-061e7a446151.

Moroccan Audit Office (2009), Audit Office Annual Report, http://www.courdescomptes.ma/upload/MoDUle_20/File_20_50.pdf.

Moroccan Audit Office (2013), Audit Office Annual Report (2013), Audit Office, http://www.courdescomptes.ma/upload/MoDUle_20/File_20_166.pdf.

Conseil Ingénierie Développement et al. (2003), "Technical and infrastructure studies," Bouregreg Valley planning project, www.bouregreg.com/tiki-download_file.php?fileId=80.

El Ghomari (2015), Assessment of water policy in Morocco, presentation at the Symposium of the French Dams and Reservoirs Committee, French Dams and Reservoirs Committee, Saint Martin d'Hères, 29 January 2015.

Housing and Town Planning Institution and Law Research Group, (GRET-GRIDAUH), Presentation on town planning law in Morocco, http://www.gridauh.fr/fileadmin/gridauh/MEDIA/2010/travaux/urbanisme_sans_frontiere/3f4e364409786.pdf.

High Commission for Planning (2006), Prospects for Morocco 2030: the demographic outlook.

High Commission for Waterways and Forests and Fight against Desertification (2015), High Commission for Waterways and Forests and Fight against Desertification website, http://www.eauxetforets.gov.ma/fr/text.aspx?id=1157&uid=37.

Paris Region Institute of Urban Planning and Development (2010), Morocco opens up to the 21st century, Les cahiers de l'IAU IdF, n°154, May 2010.

Lemaizi, S. (2015), Silting up of dams: Morocco loses 1.7bn m³ of water, Les Eco, 23 April 2015, http://www.leseco.ma/decryptages/grand-angle/28805-envasement-des-barrages-le-Morocco-perd-1-7-milliard-de-m-d-eau.html, consulted on 16 February 2016.

LYDEC (2015), Catalogue of alternative techniques – Innovative and sustainable solutions for the Greater Casablanca Region, LYDEC.

Ministry of Water (2015), Assessment of the Implementation of the National Flood Protection Plan.

Ministry of Water (no date), presentation of the National Flood Protection Plan and the impactof protective structures on the environment, www.emwis.org/documents/meetings/fol791509/fol238831/fol199761/20050428_PDI-SIG.pdf/download/1/20050428_PDI-SIG.pdf.

French Ministry of the Environment, Energy and Maritime Affairs (2013), Law of 30 July 2003: 10 years of action in the field of natural risk prevention, http://www.developpement-durable.gouv.fr/Loi-du-30-juillet-2003-10-ans-d.html, consulted on 9 April 2016.

Ministry of Housing and Town and Country Planning (2011), paraseismic regulations in Morocco–RPS 2000 Version 2011, www.mhu.gov.ma/Nouvelles%20publications/RPS2011.pdf /.

Ministry of the Interior (2015), Data sheet for the first call for projects within the framework of the Fund to Combat the Effects of Natural Disasters, Programme to Improve Resilience in Morocco when Confronted by the Risk of Natural Disasters, February 2015.

OECD (2013), *OECD Reviews of Risk Management Policies: Mexico 2013: Review of the Mexican National Civil Protection System*, OECD Publishing, Paris

OECD (2014), *Boosting Resilience through Innovative Risk Governance*, OECD Publishing, Paris.

OECD (2016), *Progress and challenges in fostering risk prevention and mitigation in a cross-country comparative perspective: Case-study report of Austria focusing on Alpine areas*, OECD Publishing, Paris.

Omira, R., M.A. Baptista, F. Leone, et al. (2013), "Performance of coastal sea-defence infrastructure at El Jadida (Morocco) against tsunami threat: lessons learned from the Japanese 11 March 2011 tsunami", *Natural Hazards and Earth System Sciences*, vol. 13, p. 1779-1794, http://www.nat-hazards-earth-syst-sci.net/13/1779/2013/nhess-13-1779-2013.pdf.

Queensland Flood Commission of Inquiry (2012), Final report, www.floodcommission.qld.gov.au/__data/assets/pdf_file/0007/11698/QFCI-Final-Report-March-2012.pdf.

Taza-Al Hoceima-Taounate Wilaya, Regional Council of Taza-Al Hoceima-Taounate, Ministry of Housing and Town and Country Planning (2010), "Taza-Al Hoceima-Taounate Regional Urban Development Plan ", Section 1 – Environment, Resources and Cultural Heritage – Natural and Technological Risks, INGEROP Consulting & Engineering, SCET-SCOM (CDG Développement), March 2010

Zurich (2015), Moroccan floods in 2014: what can we learn from Guelmim and Sidi Ifni?,https://www.zurich.com/_/media/dbe/corporate/docs/corporate-responsibility/risk-nexus-morocco-floods-of-2014-november-2015.pdf?la=en

Chapter 5.

Emergency management in Morocco

Effective major risk management can mitigate the negative human, social, economic and environmental consequences of disasters. Well-planned and organised emergency management is a critical element of disaster response. This chapter reviews Morocco's progress in emergency management while confronting serious natural disaster risks. It examines the organisational structures implemented for planning disaster response by the various agencies, warning mechanisms, inter-agency coordination and crisis communications. In order to respond to major crises, a unified crisis communication system, the involvement of civil society, and improved international cooperation are key issues to achieve further progress in Morocco.

Introduction

Over recent decades, several OECD countries have invested in improving their preparation for and response to emergencies. The development of early warning systems and emergency plans, the increased capacity of civil defence and volunteer organisations, along with better coordination, have all contributed to this improvement. These efforts resulted in a reduction in the number of victims of natural disasters in many countries. The goal of crisis management preparation is to anticipate emergencies and scale the required resources to intervene in ways that will reduce the negative effects of emergencies in an effective and coordinated manner. Even if unique events that result in a large number of victims are statistically uncommon, they must be prepared for. One of the major challenges of civil defence, beyond executing predetermined response plans, is to continuously improve the ability to anticipate unlikely events and prepare for the unexpected.

Emergency management in Morocco has a clear decision-making structure, with inter-agency coordination and ramp-up ability, coordinated centrally as well as by *walis* of various regions and provincial and prefect governors. Since the Al Hoceima earthquake of 2004, this system has significantly improved in dealing with large-scale crises based on the gaps and deficiencies identified during that disaster. Since 2007, the Ministry of the Interior's Monitoring and Coordination Centre has taken on the role of response coordinator at the national level. Following the significant flooding affecting large areas of the country, particularly in the Gharb in 2009, the winter rains of 2014 tested the system's effectiveness in coordinating response in remote areas. Despite warning and the mobilisation of important resources, the large number of victims caused by this disaster indicates additional efforts are needed to improve emergency response on the ground.

Emergency response planning

Ensuring the right response to the crisis at hand involves solid preparation. The first goal is to define the required resources and operational capacities for confronting major risks identified in the risk analysis. Following well-established procedures, regularly scheduled exercises must be included to test the plans and ability of emergency response network members to implement them when developing emergency plans that mobilise operational capacities.

Planning how to respond to disasters

In most countries emergency response capabilities are shared among several institutions. From security forces to health services, including police or the military, all the way to response units in the transportation, energy, water or telecommunications sectors, every organisation can use its capacity to contribute to the overall response in its own way (based on the nature of the crisis, its institutional structure and mandate).

The goal of planning how to respond to disasters is to ensure these entities have the necessary resources to be deployed across the country in order respond to emergencies in a coordinated manner.

Moroccan civil defence, the main actor in emergency response

Civil defence is the main actor in Morocco's emergency response. Created in 1955 by royal *Dahir*, civil defence is responsible for implementing and coordinating aid in significant disasters (Article 1). According to Decree 2-08-159 of 16 January 2009, modifying and supplementing decree 2097-176 of 15 December 1997, relating to the scope and organisation of the Ministry of the Interior, the Directorate-General of Civil Protection (DGPC) is placed under the authority of the Ministry of the Interior. DGPC is an organisation of reflection, research, study, consultation and intervention for the defence of individuals and their property in all circumstances. Its responsibilities include organising, leading and coordinating protection and aid in disasters, as well as promoting risk prevention. Civil defence is decentralised to the regional and provincial level.

To meet its responsibilities, DGPC has developed a Master Plan for Risk Analysis and Coverage (SDACR), which requires response resources to be made available at the regional level based on risk analysis (Chapter 3). DGPC's actions are based on an array of historical statistical data and include worst-case scenarios in its analysis. This analysis provides the basis for improving resource availability in the kingdom's regions, with the creation of regional intervention units, a national unit based in Salé and six regional equipment and resource depots. Four additional depots are currently under construction. DGPC's progressive resource building is supported by the Special Fund for the Promotion and Support of Civil Defence, established in 1997 and funded by payment for services rendered by civil defence. On average, an annual 250 million Dirhams are expended from this fund, 53% of which is spent on equipment (Table 5.1).

Table 5.1. Expenditure of the Special Fund for the Promotion and Support of Civil Defence in millions of Dirhams

Areas	2009	2010	2011	2012	2013	2014 (est.)	2015 (est.)
Acquisition of vehicles, aid, field, technical and IT supplies and equipment	-	-	88.00	37.73	123.71	12.00	20.50
Construction (barracks, etc.)	-	-	19.82	13.47	53.47	39.60	35.50
Acquisition of clothing items	-	-	18.60	21.38	53.00	48.20	50.20
Vehicle depot operation	-	-	-	26.00	87.00	36.00	30.00
Night shift compensation	-	-	46.00	37.41	133.70	46.00	46.00
Other	-	-	31.18	3.00	36.37	0.90	11.50
Total expenditure	**352.28**	**192.39**	**203.60**	**139.00**	**487.25**	**182.90**	**193.70**

Source: Report on Special Treasury Accounts 2013, 2014 and 2015, Moroccan Ministry of Economy and Finance

Given Morocco's risk profile, positioning resources across the country while strategically locating more effective national resources, for example, the INSARAG certified civil defence unit (Box 5.1), seems to be a good approach. However, with a total of 8,000 staff members, civil defence has one agent for every 4,000 inhabitants, which is still low in comparison to OECD countries, where ratios higher than 1 to 1,000 are the norm. Concerning material resources, civil defence has one emergency vehicle for every

34,000 inhabitants (Ministry of the Interior, 2005), which seems insufficient for dealing with a very large-scale disaster. The country's extensive geographic area is another challenge in the proper distribution of resources. Resources are concentrated in the country's northern half, which is more densely populated and where disasters (floods, earthquakes and forest fires) are more frequent. Therefore, providing assistance to more isolated provinces means aid can be delayed because of the need to cover greater distances. Floods in the south of the country during the end of 2014 demonstrated the need to make more resources available in the southern part of the country.

Box 5.1. A Civil Defence unit receives INSARAG certification

In October 2014, Morocco became the first African country to have one of its Civil Defence units fulfil the criteria established by the International Search and Rescue Advisory Group (INSARAG) and be certified as an Urban Search and Rescue team in the "heavy" classification. This certification means the unit will be part of international activities of civil defence forces under the guidance of the United Nations. It will also facilitate foreign assistance in the event of a large-scale disaster on Moroccan soil.

A Moroccan Civil Defence detachment from Salé, including close to 120 people, successfully participated in a mock earthquake exercise in October 2014 in Epeisses, near Geneva. This specialised unit is comprised of a search and rescue operations management team, four rubble rescue teams, four technical search teams, four canine teams, four medical teams and an operational logistics technical team. "Heavy" classification USAR teams have an operational ability that allows them to conduct complex search and rescue operations, especially in urban environments with collapsed or weakened structures. They must provide the equipment and personnel necessary to intervene continuously in two distinct locations for 10 days, domestically or internationally. This certification represents an achievement of Morocco's ongoing efforts to strengthen its Civil Defence capacity, including with the support of Swiss Cooperation, since 2008. This has involved 206 training sessions on varied subjects including crisis management, search, rescue, logistics and medicine.

INSARAG is a worldwide network that brings together more than 80 countries under the guidance of the United Nations. INSARAG deals with issues of search and rescue in urban environments. Its goal is to establish international standards to which urban search and rescue teams must adhere. INSARAG also works for improved coordination of international response to earthquakes, according to INSARAG guidelines approved in 2002 by United Nations Resolution 57/150 on "Strengthening the effectiveness and coordination of international urban search and rescue assistance."

Sources: Interviews conducted by the OECD for the Study on Risk Management in Morocco (May 2015); INSARAG (2012), *INSARAG External Classification/Reclassification Handbook, 2012 Edition;* MAP (2014), "Disaster Management – Morocco Becomes First African Country to Meet UN Standards," Le Matin, 29 October 2014, http://www.lematin.ma/express/2014/gestion-des-catastrophes_le-Morocco-premier-pays-africain-a-se-mettre-aux-standards-de-l-onu/211358.html, reviewed 2 November 2015.

Contributions from other actors in the emergency response network

Beyond civil defence resources, emergency response requires the mobilisation of an entire range of sectors within the emergency response network. The first line consists of health, security and infrastructure players.

It is often necessary to mobilise the human and material resources of the Royal Armed Forces and other security agencies to supplement civil defence resources in major crises, but the process is not well formalised. The close-knit nature of the country and the ability to deploy its forces rapidly is an advantage in crisis response. The armed forces are often the only institution with specific capabilities required to respond to large-scale disasters (the Royal Gendarmerie's helicopters, the Royal Armed Forces' field hospitals). During the 2004 Al Hoceima earthquake, significant resources, particularly contingents from the army, navy and gendarmerie, were deployed on an emergency basis to supplement rescue operations conducted by civil defence. However, the definition of operational resources available in natural disasters is not formalised and is not based on the civil defence risk assessment or an evaluation shared among the various players nationally. It also does not provide a clear understanding of their contribution to crisis management.

Health services in Morocco are currently being strengthened, particularly since the adoption of the National Strategy for the Management of Medical Emergency and Health Hazards along with Disaster Events in 2005 and the more recent Medical Emergency Action Plan of 2012-2016. The objective of this strategy is to strengthen emergency services in Morocco's hospitals, ambulances, helicopter transport and development of emergency medical services through a 500 million Dirhams investment. It also requires all hospitals to develop a hospital emergency plan, including identifying risks for disasters that may affect the hospital's service area or the hospital itself. Although this capacity building is based on statistics indicating an increasing need for emergency care for the population as a whole, it does not seem to specifically address major disasters. It will contribute significantly to improving management of injuries during disasters by building closer links across the country. However, it is not based on a prospective needs evaluation for risks of large-scale disasters. The Al Hoceima earthquake demonstrated the need for capacity in regional hospitals to deal with large numbers of patients during major disasters.

The critical infrastructure sector also possesses response resources that can contribute to disaster management. The Ministry of Public Works, Transportation and Logistics and its road management function deploys technical teams across the country to provide transport route security and rapid repair during disasters. Water, electricity and telecommunications systems operators, ONEE as well as private firms, all have response teams to provide continuity of critical services (Box 5.2). Some are very well equipped for this, but they plan their resources independently of civil defence and based on their own evaluations in each different area. The number of diverse initiatives makes it difficult to precisely evaluate the adequacy of the resources needed to respond to a potential crisis.

Box 5.2. Emergency response by a critical infrastructure operator: Lydec's approach in Casablanca

Under a 30-year management subcontract signed in 1997, Lydec manages water and electricity distribution, wastewater collection and treatment, rainwater collection and public lighting for the Greater Casablanca (Morocco) Region. During the rainy season, it is responsible for coping with emergency situations. Casablanca regularly experiences flooding. In January 1996, the city received 125mm of rainfall, which resulted in many deaths.

To plan for major climatic events and mobilise its response resources effectively, Lydec signed an agreement on 1 July 2010 with the National Meteorological Office (DMN) for customised meteorological forecasts. This partnership involves the prevision by DMN of meteorological forecasts for the Greater Casablanca Region in the form of daily bulletins and special bulletins in extreme weather situations. It also stipulates that radar animations are to be transmitted by the DMN and training in the use of meteorological information should be provided to Lydec associates. Based on information provided by the DMN, Lydec scales its response on three "rainy season" alert levels:

- Yellow alert: "Mobilisation": heavy rain forecast within the next 48 hours, response teams on alert.
- Orange alert: "Deployment": heavy rain confirmed, teams deployed on the ground (fixed observation or monitoring rounds) and ready to respond to facilities if needed.
- Red alert: "Response": heavy rain in progress, teams activated and moved based on the event's evolution.

This partnership will assist Lydec in emergency management and in fulfilling the ongoing requirement for operational monitoring through its "Central Management Office." It allows for continual observation and remote response across all networks and facilities in Casablanca. To strengthen their response ability, Lydec teams use a modern geolocation system and waterproof 3-4G Wi-Fi video equipment accessed via Android tablets or smartphones to identify functional obstacles in the sanitation system. They can also transmit strategic information and provide images of the intervention area to the central crisis management platform. It is important to note these tools have not yet been used during times of natural disasters. They are, however, used on a regular basis to monitor the systems' operations.

Source: LYDEC (2014), A "Smart City" Approach for Greater Casablanca: Stakes, Actions & Opportunities, presentation in the Smart Cities programme, "The City of Tomorrow Project", LYDEC, 23 September 2014, Casablanca, http://www.cgem.ma/upload/1391890939.pdf.

Overall, while resources are adequate to respond to small or medium impact disasters, they appear limited when compared to anticipated needs. Therefore, it seems important to ensure their optimal utilisation and appropriately channel resources to the serious risks threatening Morocco. Although civil defence plans are based on risks in conjunction with the SDACR for the use of emergency resources, there is no centralised crisis management needs evaluation process that records all available resources to cope with large-scale crises. These steps are crucial to identify potential gaps and planning for more efficient resource allocation. Examples of capacity planning implemented in Britain and the Netherlands may represent a source of inspiration for Morocco (Box 5.3).

Box 5.3. Resource planning process in European Union countries

In the Netherlands and United Kingdom, response resource planning is based on a national multi-agency risk assessment process that assembles all of the reasonable worst case scenarios and evaluates their potential impacts and the probability of their occurrence. Following this exercise, which receives overall approval from the central government, each emergency response player must specify the resources it assigns to respond to these scenarios within its own areas of responsibility. This provides consistency across emergency management systems in equipment, human resources and emergency supplies. This approach has since been adopted at the European level and all European Union countries are now developing a similar approach.

Source: OECD (2015), *The Changing Face of Strategic Crisis Management*, OECD Publishing, Paris.

Multi-player response plans and sector response plans

Once emergency response operational capacities are established, emergency response plans must be developed to mobilise these resources effectively when a crisis occurs. In Morocco, crisis response management includes a standardised multi-player plan (ORSEC), extended locally within regions under the authority of the *walis* and governors, with specific plans for specific risks. However, the development of emergency plans is uneven across the country, and plans are not yet regularly updated.

The ORSEC plan, the basis of Morocco's emergency response

The ORSEC "Emergency Organisation" plan, implemented in the 1980s, provides the generic structure for regional and provincial emergency response in Morocco. It can be activated by the *wali* or governor of the *préfecture* in the event of a large-scale disaster. In accordance with current laws and regulations, ORSEC plans for the requisition of the resources and persons necessary to manage the crisis under its authority. Its activation leads to the establishment of a crisis unit in the wilaya or the *prefecture* articulated around six primary missions that mobilise key actors of various action areas (Table 5.2). The ORSEC plan connects important players involved in emergency management and coordinates their response on a territorial level.

Table 5.2. Organisations taking part in the ORSEC plan

Missions	Organisations involved
Liaison and transmission	Telephone and cell phone transmission services
Maintaining order	Police, gendarmerie
Search and rescue	Civil Defence, fire brigades, Royal Armed Forces, auxiliary forces
Medical care	Ministry of Health, Moroccan Red Crescent
Construction and equipment	Ministry of Public Works, Transportation and Logistics, infrastructure operators (National Electricity and Potable Water Office, etc.)
Food and shelter	Civil Defence, Moroccan Red Crescent, associations

Source: Khaldi (The quality of ORSEC plans differs by region. The information included in these plans, how up-to-date they are, and the scope of risks covered is uneven. However, examples of highly developed operational plans can be found in some areas. In the Souss-Massa Dra Region, for example, the ORSEC plan describes in detail the roles and resources of each response actor and specifically identifies vulnerabilities to various risks. In other cases, the ORSEC plan is limited to a list of institutions and contacts along with administrative tasks for each in a crisis situation. The Tangier-Tétouan Region represents an example of this type of plan. These differences indicate significant heterogeneity in the ability to manage large scale crises and provide an effective response. There does not seem to be a standardised methodology across regions and provinces for developing individual ORSEC plans.

Sector plans

In addition to the multi-player ORSEC plan, agencies involved in emergency management have implemented specific emergency plans within their particular sector. These plans are quite heterogeneous, with some action plans meeting the highest international standards, while others provide very sparse details. Moroccan Civil Defence has also developed operational plans to mobilise its resources for several different risks, such as the SINON flood plan or the SEFER railway accident plan. In the health sector, the "Red Plan" is triggered by the governor in order to activate health resources for the evacuation of large numbers of casualties, and "White Plans" are developed by hospitals to provide triage and accommodation of the injured.

In critical sectors, the approach of METL, its agencies and public or private water management operators, has demonstrated awareness of the importance of continuity in these sectors. The Ministry of Public Works, Transportation and Logistics has chosen to plan its own resources to ensure the continuity or rapid restoration of the critical infrastructure under its responsibility. This planning includes launching reconnaissance operations to determine the state of critical points in the system and monitoring or verifying outages, allocating additional human resources, deploying signage and diversion equipment as well as performing temporary construction on damaged sections of infrastructure (Box 5.4). ONEE has also implemented detailed emergency plans, consisting of "reaction" files to cope with various types of damage to the water infrastructure for which it is responsible. These standardised procedures are tailored to each region, and are an example of the most advanced emergency plans currently available. Private water companies, such as LYDEC in Casablanca, have also developed emergency plans, as have the Hydraulic Basin Agencies, who have emergency plans to adapt to the management of major dams and to mobilise their teams in case of flooding. Finally, the National Railways Authority (ONCF) has also developed emergency plans for certain disasters that may impact rail services, such as flooding.

Box 5.4. Multi-agency crisis management approaches of the Ministry of Public Works, Transportation and Logistics

The Ministry of Public Works, Transportation and Logistics (METL), tasked with government policy in the areas of roads, ports, maritime, air and rail, implemented a dedicated crisis management structure in 2010 known as the High-Risk Situation Management Division. This agency is tasked with leading and coordinating the response among the Ministry's various operational agencies, as well as with the different organisations under its leadership, such as the National Railways Authority (ONCF), the National Airports Authority (ONDA), the National Ports Authority (ANP), and the Moroccan National Highway Corporation.

The High-Risk Situation Management Division provides regular follow-up on warnings related to weather, informs central government services about the state of infrastructure on a real-time basis, and provides communications on power cuts to various agencies. If infrastructure is impacted, it can efficiently become involved in notifying and organising contingencies, and then carry out temporary work to restore that infrastructure. For flood risk to the highway network, the division uses an early warning system called "Inforoute," which allows it to gather information in real time on possible incidents and share it among the Ministry and its related organisations so that an intervention can take place.

The Division is also part of the larger coordination approach organised by the CVC: it is a member of this structure, where it represents the Ministry in the CVC crisis group for emergencies.

Sources: Ministry of Public Works, Transportation and Logistics (2015), Prevention and Management of Flood Risk within the METL, presentation for the OECD Review of risk management in Morocco, 6 June 2015, Rabat; KPMG (2013) Report on Definitions, Current State of Affairs and Diagnosis, study for improvement of the current crisis management system for the Ministry of Public Works, Transportation and Logistics, KPMG.

Scenario-based plans

Individual response plans, specific to certain risks, have also been developed. For example, the Souss Massa Draa region has an individual "flood" plan in addition to the El Haouz' provincial plan. These plans identify specific risk points and actions to be implemented in a disaster. However, these plans are not in place in all regions, not updated regularly, insufficiently tested, and they rarely identify specific vulnerabilities. In national as well as regional plans, not all major risks are covered by specific emergency plans. For example, there is not yet a plan for the risk of tsunamis, for the scenario of a significant earthquake affecting a large part of northern Morocco, or for the risk of two large simultaneous floods in two remote areas of the country. Planning for emergency response adds increased value in anticipating these types of risk. As a result, it also changes the level of vulnerability to risk. As such, the city of Agadir launched the development of an emergency response plan to a large earthquake as part of a project that was financed by an FLCN request for proposals (see Chapter 4).

In addition, a specific emergency plan for drought would be beneficial when reduced rainfall is detected. This is particularly important given the possible effects on the agriculture sector and on Morocco's economy. Faced with a rainfall deficit during the 2015-2016 agricultural season, the Moroccan government provided an emergency allocation of 4.5bn Dirhams to protect plant resources, specifically for irrigating some rain-fed plantations, securing seeds and water resources for the following season, providing drinkable water to affected villages, and protecting animal resources by providing feed and water for livestock.

Drills

To ensure emergency plans are operational, improving knowledge of procedures and identifying potential failures, testing emergency plans and procedures using drills are key components of the crisis preparation phase. In Morocco, simulation exercises are used to test strategic, tactical and operational approaches, including working with international partners. According to some emergency response actors, crisis drills in Morocco are not held frequently enough to achieve their goals. Operationally, several players conduct crisis management drills focused on various issues and with multiple partners, with the private sector or internationally. For example, ONCF has worked with the Red Crescent on railway accidents, ONEE regularly tests its procedures, and LYDEC has simulated large-scale electric blackouts with Civil Defence. Ports and users of port facilities organise emergency security exercises in support of the National Ports Agency's emergency plan. Every two years, a national exercise is conducted simulating a marine petroleum spill, called "Simulex". These drills have improved emergency procedures.

At the strategic level, drills are not conducted systematically, although they are essential to improve crisis coordination and decision-making. Simulations have been conducted with international partners, such as the "Westsunami 2015" tsunami warning drills in partnership with Spain and Portugal in 2015. This type of strategic coordination exercise is regularly conducted by OECD member countries, gathering high-level decision makers to develop their crisis management abilities. In Germany, strategic crisis management exercises involve cabinet ministers and Länder presidents every two years based on national crisis scenarios (Box 5.5).

Box 5.5. A strategic crisis coordination exercise: LUKEX in Germany

LÜKEX is a national strategic crisis management exercise led by the Ministry of the Interior and the German Federal Office of Civil Protection and Disaster Assistance (BBK).

LÜKEX is a comprehensive exercise conducted at the strategic level (convening inter-ministerial crisis units, political and administrative teams, inter-sector coordination teams, etc.). The exercise has taken place on a biannual basis in Germany since 2004. The aim of the exercise has been to raise players' awareness of the primary problems in crisis management and test their ability to react to major risks. Each exercise involves the participation of institutions at the federal level. For LÜKEX 15 (a storm surge scenario), the Ministries of the Interior, Transport, Defence at both centralised and decentralised levels, the national meteorological service and civil defence all participated. The exercise created both horizontal and vertical cooperation within the government and with critical infrastructure operators, in areas such as healthcare, energy and media. Each crisis simulation (scenario) involves a 24-month cycle including planning (6-8 months), preparation (9-11 months), execution (2-3 months) and evaluation (4-5 months), creating connections between decision makers prior to the actual exercise. The exercise itself is planned to be as comprehensive as possible, and integrates strategic and operational activities by having the crisis unit function under real working conditions.

Source: OECD (2016), Toolkit on Risk Governance, https://www.oecd.org/governance/toolkit-on-risk-governance/, consulted on 12 February 2016.

Although some multi-actor drills have been conducted in Morocco, there have not been a sufficient number of exercises that convene all national crisis management actors. Exercises to familiarise the population with life-saving actions to be taken in a high-risk situation have also been lacking. For example, no drills were conducted on the implementation of the flood warning system in the Ourika Valley or on the risk of earthquake. It could be relatively simple to organise and conduct simulation exercises for major earthquakes in order to reduce their potential cost to human life, as it was demonstrated by the American "Shake Out" exercise (Box 5.6).

Box 5.6. The Shake Out earthquake drill conducted in the United States

The goal of the Shake Out drill is to inform the public how to prepare and what to do during a major earthquake. The exercise promotes the idea of "drop, cover and hold on", reducing the risk of injury and death in a large earthquake. Started in southern California in 2008, this exercise is held every year in more than 20 regions internationally (including Japan, New Zealand and Canada). In 2015, 22 million people participated around the world.

The Shake Out exercise is organised by the Earthquake Country Alliance, a partnership of national, federal, local and civic organisations. The Alliance includes: The California Emergency Management Agency, the U.S. Geological Survey, the California Earthquake Authority, the American Red Cross, as well as the Southern California Earthquake Centre, which directs the Alliance. Initially, the purpose of the Shake Out exercise was to prepare southern California for the "Shake Out" scenario, which modelled a magnitude 7.8 earthquake on the San Andreas fault. Individuals, businesses, schools, religious groups, community groups, government agencies and other organisations can sign up to participate on the Shake Out website.

Source: OECD (2016), Toolkit on Risk Governance, https://www.oecd.org/governance/toolkit-on-risk-governance/, consulted on 12 February 2016.

Activation and implementation of emergency response

When a disaster risk is detected, emergency response mechanisms can be activated and rescue operations can begin. Response may begin when a risk is anticipated, by triggering warning systems already in place for detecting disaster risks and forecasting their evolution,. Response to unforeseen events may also begin in real time, when emergency services are notified by the initial witnesses or victims. Implementing an emergency response starts with an overall system ramp-up based on the seriousness of the crisis and information received on the ground. The response's effectiveness depends on how rapidly the disaster is forecast or detected, and on good coordination with effective communication between the players. The goal is to mobilise intervention resources as quickly as possible to disaster sites, and to ensure they are fully responding to the need.

Effectiveness of Morocco's early disaster warning systems

Activation of emergency plans by early warning systems has progressed significantly over the past 20 years, and has saved many lives around the world. In Morocco, high-performance warning systems have developed rapidly over the last few years. By allowing vulnerable populations to take the necessary measures to protect themselves and government agencies to mobilise resources, these systems are important tools to limit the impact of potential risks. Their success depends on the effectiveness of systems those measures, track and forecast hazards, as well as the ability to transform data into viable and intelligible warning messages to different segments of the population (urban/rural, tourists, immigrants) or to businesses. Their effectiveness also depends upon the public's understanding of what decisions to take in an emergency situation. A successful warning system must fully connect authorities with the affected population (Box 5.7).

Box 5.7. Ten rules for an effective early warning system

According to the World Meteorological Organisation, applying ten key rules can ensure the effectiveness of early warning systems (EWSs) and reduce human and economic losses from hazards. Based on an analysis of various EWSs in developed and developing countries, as well as lessons learned from implementing these systems, these rules are universal and independent of the socio-economic, political, cultural and institutional characteristics of a specific country or context.

1. Strong political support is essential to an early warning system's success: There is a strong political recognition of the benefits of EWSs reflected in coordinated national to local disaster risk management policies, planning, legislation and budgeting.
2. Operational detection and warning capacities provided by EWSs must be accompanied by improved community preparation: Effective EWSs are built on four components: (i) hazard detection, monitoring and forecasting; (ii) analysing risks and incorporating risk information in emergency planning and warnings; (iii) disseminating timely and "authoritative" warnings; and (iv) community planning and preparedness.
3. EWS roles and responsibilities must be clear: EWS players are identified and their roles and responsibilities and coordination mechanisms clearly defined and documented within national to local plans, legislation and directives.
4. The EWS must receive adequate resources: EWS capacities are supported by adequate resources (e.g., human, financial, equipment, etc.) to ensure its proper functioning.
5. Information on risks must be used to improve emergency and warning systems: Hazard, exposure and vulnerability information are used to conduct risk assessments at different levels, as critical input into emergency planning and development of warning messages.
6. Warning messages must be effective, "authoritative" and concrete: Warning messages: (i) are clear, consistent and include risk information; (ii) link threat levels to emergency preparedness and response actions (e.g., using colour codes, instructional images, etc.); and (iii) are issued by a single, recognised and "authoritative" source.
7. Warning messages must be sent to involved players in a timely manner to allow them to react: Warning messages are able to reach the authorities, other EWS players and the public at risk in a timely and reliable fashion.
8. Emphasis must be placed on preparation and planning prior to the emergency: Emergency response plans are developed with consideration of hazard/risk levels, characteristics of the exposed communities (for example, rural, urban, tourists, most vulnerable populations, etc.) and include coordination mechanisms between the various EWS players.
9. Risk training and drills are essential for operational preparation: Training in hazard/risk/emergency preparedness awareness is integrated into various formal and informal educational programmes and information is provided on hazard detection and steps to take in emergency situations. Additionally, regular drills must be held to ensure EWS operational readiness.
10. Feedback and continual improvement must be an ongoing priority: Effective feedback and improvement mechanisms are in place at all levels of EWS to provide systematic evaluation and ensure system improvement over time.

Source: Golnaragui M. ed. (2012), *Institutional Partnerships in Multi-Hazard Early Warning Systems*, Springer-Verlag Berlin Heidelberg.

Hydro meteorological risk warning

Building capacity in the National Meteorological Office (DMN) has significantly increased Morocco's warning abilities. In 2010, FLCN invested 135.5 million Dirhams in the DMN's triennial action plan, including an expansion in meteorological radar coverage, strengthening of the automated observation network, improvement in digital forecast models, development of data collection systems and dissemination of warning bulletins (Ministry of Economy and Finance, 2015). Now, DMN has the technical resources and skills in refined meteorological forecasting to improve the quality and timeliness of meteorological warnings and to better mobilise emergency response forces ahead of crises. In dangerous weather conditions (particularly heavy rainfall or heat waves), a pre-alert bulletin is developed 12 to 24 hours in advance during the winter season, and 6 to 12 hours in advance for summer storms. During the pre-alert period and following heightened monitoring of the situation, forecasters can confirm or discontinue the pre-alert bulletin. Warning bulletins are communicated directly to the Ministry of the Interior and provincial governors, who are responsible for informing the local authorities, and to a pre-established list of users with DMN agreements, particularly in critical infrastructure (Ministry of Equipment, ONCF, ANP, Lydec, etc.).

Regarding flood monitoring, a partnership between DMN and the Hydraulic Basin Agencies (ABH) follows the hydro meteorological status of watersheds. DMN heavy rain warnings include the implementation of a pre-alert system and close monitoring of changes in waterways by ABH teams. The flood alert network is based on a system of HF radios (225, according to DMN) dispersed within ABH watersheds and centralised in the Rabat water management agency. According to numerous stakeholders, this system would benefit from being improved, better equipped and automated. Although it is used to manage dams, it is used less frequently to warn populations in imminent danger. Only the Flood Prevention and Warning System (FPWS) established in the Ourika Valley in 2001, seems to be truly effective in providing specific advance flood warnings that allow the evacuation of the valley during heavy flooding. In this steep-sided basin where flash floods can go from a rate of 10 to 1000 sq. metres per second in less than a half-hour, with a particularly deadly result in 1995, the FPWS played a spectacular role in detection and warning during ten floods between 2003 and 2013 (Box 5.8). Within the National Flood Protection Plan (PNPI), other water resource measurement and monitoring systems were launched in several watersheds, including Loukkos, Oum Erbia and Moulouya (Minister for Water, 2015).

Box 5.8. Early Warning System for floods in the Ourika Valley

A major tourist site, the Ourika Valley was hit by violent flooding on 17 August 1995, resulting in 242 deaths, affecting 35,000 people and causing over USD 9 million in damage. These floods significantly slowed the area's development plans during the 1990s. To limit potential human and economic losses when flooding occurs again, local authorities decided to install a flood warning system in the Ourika Valley.

With the assistance of the Japanese International Cooperation Agency (JICA), the Ourika Valley was provided with a Flood Prevention and Warning System (FPWS) in 2001, consisting of 5 observation stations, 2 data transmission stations, 4 monitoring posts and a warning post located at Ighref. Since 2007, the Hydraulic Basin Agency (ABH) in Tensift has also taken part

Box 5.8. Early Warning System for floods in the Ourika Valley *(continued)*

in this project by financing the extension of the telemetry network to 16 sites, including 3 pluviometric and 13 pluviometric and limnometric sites. In collaboration with JICA, the Ourika Valley FPWS is currently part of an extension project to equip six hydrologic stations with automatic measurement equipment provided by JICS (*Japan International Cooperation System*).

The Ourika Valley FPWS played a successful part in detection and warning for 10 floods between 2003 and 2012. During the 2014 floods affecting Al Haouz Province no loss of life occurred despite some property damage. In addition to providing an efficient early warning system, this project connected local populations with emergency procedures. Today, when flooding occurs, many local volunteers are tasked with warning tourists and helping them evacuate the areas at-risk.

Source: Interviews conducted by the OECD for the Study on Risk Management in Morocco (May 2015); Tensift Hydraulic Basin Agency (2015), "Management of Risks Associated with Floods – Early Warning System for Floods in Tensift Basins – Al Haouz Area Case," presentation during meetings conducted by the OECD Secretariat during the Peer Review, Tahannaout, 2015.

Despite their strengthening over recent years, it is important to note these warning systems provide uneven coverage. For floods, only 60% of watersheds are equipped with a water height and rate telemetry system, even though very few real-time models can transform weather predictions into flood predictions. PNPI includes the development of a master plan for the implementation of nationwide telemetry systems. The cutting-edge technology put in place for the Ourika Valley automated flood warning system should inspire improvement for other Hydraulic Basin Agencies.

DMN warning systems could also be simplified by using colour codes or covering a greater number of risks. Work is still required to ensure warning messages are adapted to the needs of local populations, and messages are communicated more directly. While some partnerships formed with ONCF or LYDEC have allowed them to receive weather warnings and activate emergency plans, in other cases, DMN warnings have not always led to appropriate risk management, as in the case of the Guelmim floods (Box 5.9).

Box 5.9. The Guelmim floods

In November 2014, in Guelmim and Sidi Ifni, heavy rain caused serious flooding, resulting in the death of 47 people and significant property damage. This rain was the subject of warning bulletins sent by the National Meteorological Office, but these were not enough to encourage local officials to take emergency measures to protect the population. It was only after the first devastating episode of rain the Monitoring and Coordination Centre (CVC) was activated, Civil Defence and RAF emergency services were mobilised, and the actions of local authorities were organised. Implementing a coordination structure was beneficial: it allowed the authorities to take adequate emergency measures in a coordinated way. This limited the loss of human life during a second episode of rain during the following days. The local authorities also decided prior to the flood crest to close the major roads in the at-risk area and ordered the evacuation of vulnerable areas.

Source: Zurich (2015), *Floods in Morocco in 2014: What Lessons Can Be Learned From Guelmim and Sidi Ifni?*, www.zurich.com/_/media/dbe/corporate/docs/corporate-responsibility/risk-nexus-morocco-floods-of-2014-november-2015.pdf?la=en.

Other warning systems

In Morocco, it is not possible to forewarn populations of an imminent earthquake, as has been attempted in some OECD countries (Mexico, Japan), particularly because of the distance between active seismic zones and risk areas. However, rapid earthquake detection helps emergency services to mobilise more quickly, and in some cases to automatically halt critical infrastructure to reduce potential damage. The National Geophysics Institute (ING) is responsible for transmitting earthquake warnings to the authorities. It provides 24-hour surveillance using on-call personnel who monitor the nationwide seismic telemetry network (Chapter 3). Satellite communications equipment allows warnings to be sent to 15 ministries as soon as an earthquake measuring greater than 3.5 on the Richter scale is detected.

Concerning the risk of a tsunami, Morocco does not yet have an operational national warning system. However, one is being developed under the leadership of ING. It has the infrastructure for earthquake detection and is currently developing a wave telemetry system to detect tsunamis. In addition, Morocco participates in the Northeastern Atlantic Tsunami Early Warning and Mitigation System (NEAMTWS) established by UNESCO. This regional partnership's goal is to define, in a concerted manner, a tsunami action plan. This includes a mechanism for transmitting warnings to key contact points that are constantly connected and that, in turn, communicate the warning to Civil Defence authorities. However, in the absence of a specific action plan for providing tsunami warnings and informing the public of the risk, the effectiveness of tsunami warnings in Morocco is still very limited.

Finally, while there is no specifically dedicated warning system for drought, monitoring precipitation and reservoir water levels provides information about drought risk for agriculture several months before the effects become obvious. The drought action plan, started in January 2015, also demonstrates the effectiveness of Morocco's anticipation of crisis situations. However, this anticipation can probably be improved by using the DMN's seasonal climate forecasts based on dynamic global and regional models as well as statistical models.

Ramp-up and strategic coordination

Although Morocco has implemented emergency management policies since the 1970s, particularly with the development of the ORSEC plan, the 2004 Al Hoceima earthquake revealed weaknesses. Specifically, ramp-up ability, strategic coordination and inter-agency cooperation issues demonstrated the limits of such an approach (Box 5.10). These issues led the Moroccan government to improve coordination by developing a new approach and by establishing a Monitoring and Coordination Centre (CVC) in 2008 under the leadership of the Ministry of the Interior.

Box 5.10. Improving coordination: lessons learned from the 2004 Al Hoceima earthquake

On 24 February 2004, the city of Al Hoceima and its surroundings were hit with a 6.4 magnitude earthquake that destroyed over 19,000 buildings and resulted in approximately 550 deaths.

Although local emergency services responded in the first hours of the disaster, they were quickly overwhelmed by the extent of the damage. Due to the lack of a pre-established coordination structure, obtaining support from emergency services of other regions of Morocco first, then assistance from the Royal Armed Forces, and finally international aid, took over 24 hours. Due to a lack of resources and the fact the earthquake occurred in a hard-to-reach mountainous area, the most remote villages remained isolated for several days. In addition, a lack of equipment led to delays in clearance and evacuation operations to remove victims from the rubble. Humanitarian aid to the affected populations, especially tents, tarpaulins and food supplies, took time to reach the area. These delays resulted in significant protests among the affected population, who had to rely on their own abilities to organise themselves rather than on the authorities. This, in turn, led to social unrest and demonstrations in the most affected cities in the area of Al Hoceima and Imzouren in the days following the earthquake.

These shortfalls led to the creation of the Monitoring and Coordination Centre (CVC) under the leadership of the Ministry of the Interior.

Source: Interviews conducted by the OECD for the Study on Risk Management in Morocco (May 2015); l'Économiste (2004), Al Hoceima earthquake, the town counts its dead, n°1712, 25 February 2004, http://www.leconomiste.com/article/seisme-del-hoceimabrla-ville-compte-ses-morts, reviewed 18 February 2016.

During crises, the CVC is activated at the request of the *walis* when their own regional resources are insufficient to respond and national resources are needed. The CVC takes a subsidiary approach and facilitates coordination and cooperation among the actors, playing a central role in information gathering and providing help in decision making and response coordination. It ensures the allocation of human and financial resources during the crisis, and guarantees communications. It has a crisis room that brings representatives of the various crisis management agencies together in Rabat, particularly Civil Defence, the Royal Gendarmerie, the Royal Armed Forces, the Ministry of Public Works, Transportation and Logistics, and the Minister for Water. This approach gathers information from the field, with each of these organisations activating its internal crisis management processes relatively independently (Figures 5.1 and 5.2).

Figure 5.1. Organisational structure of the Monitoring and Coordination Centre

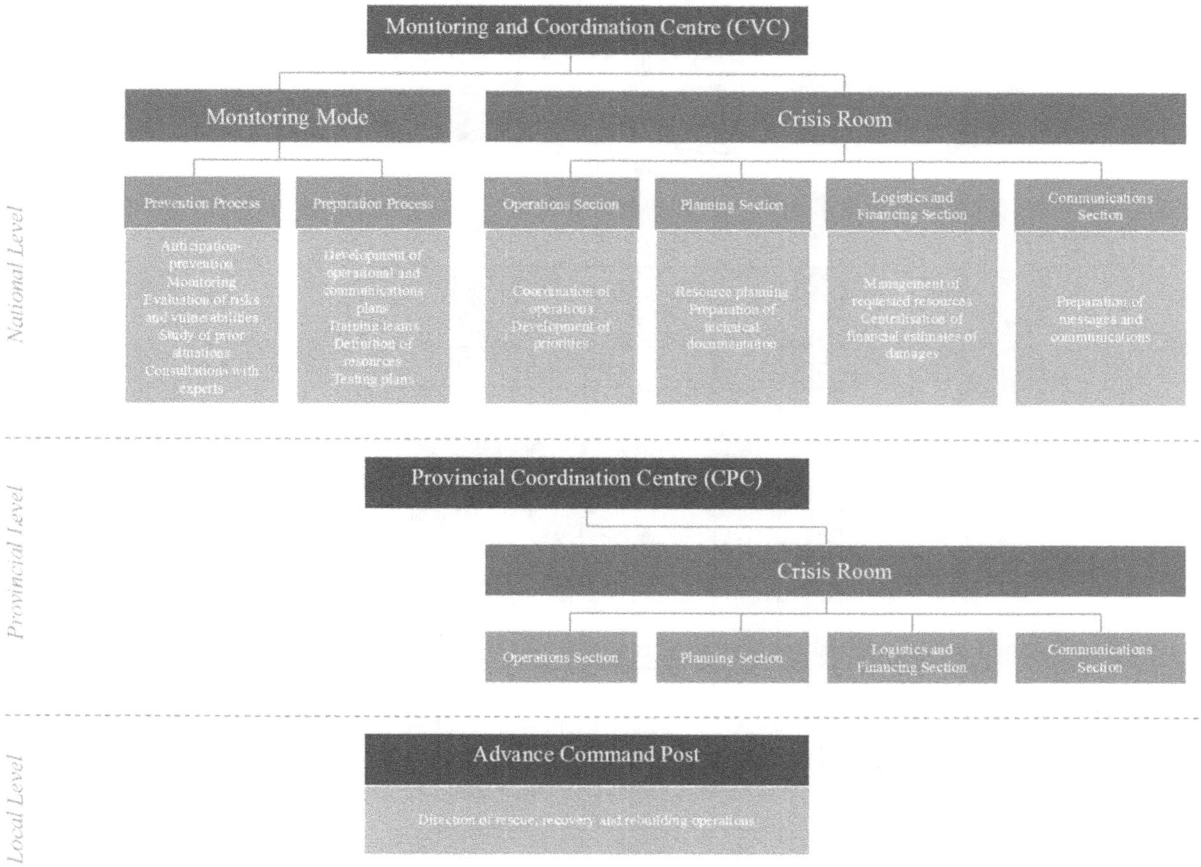

Source: Interviews conducted by the OECD for the Study on Risk Management in Morocco (May 2015); Monitoring and Coordination Centre (2014), Presentation on the New Approach to Crisis Management, presented to the Ministry of the Interior, 20 February 2014, Rabat.

Figure 5.2. Formal approach to crisis situation ramp-up

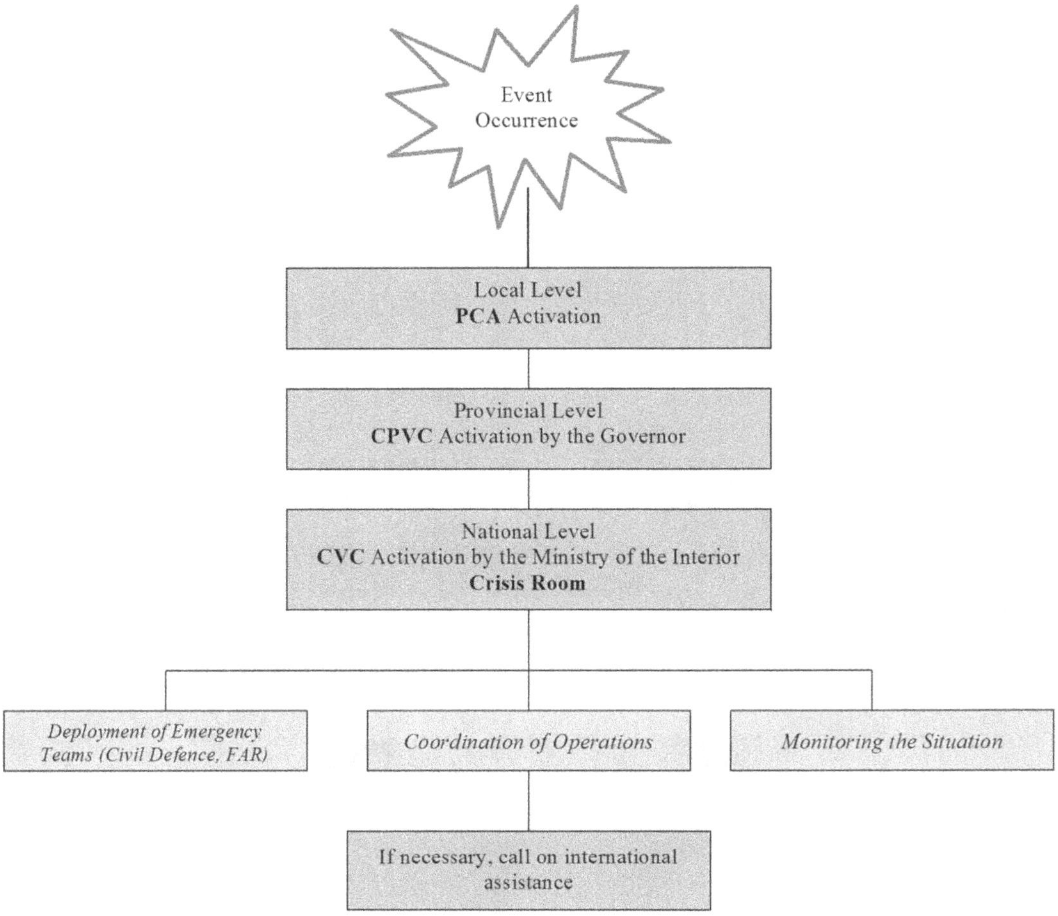

Source: Interviews conducted by OECD for the Study on Risk Management in Morocco (May 2015); Monitoring and Coordination Centre (2014), Presentation on the New Approach to Crisis Management, presented to the Ministry of the Interior, 20 February 2014, Rabat.

While the establishment of the CVC and its replication at the level of the *wilaya* has contributed to improve crisis coordination and mobilisation of additional resources when required, operational coordination on the ground still suffers from a fragmented approach between different sectors. The establishment of Advanced Command Posts (PCA) or Area Command Posts (PCZ) in multi-site crises is a common practice that can make use of Civil Defence resources, particularly with the mobile command post available in Casablanca. However, there is no formal coordinator for multi-agency emergency services on the ground, and each organisation works independently within its own area or taking an *ad hoc* approach to each crisis.

The effectiveness of coordination and ramp-up mechanisms during crises is also based on how quickly decisions can be made and the flexibility to confront unforeseen situations. The activation of the CVC and crisis units in *wilayas* and prefectures is an essential part of being able to act in an emergency. The implementation of these units is still too dependent upon decisions made by the Ministry of the Interior and the *walis*. It would benefit from more automatic activation based on warnings from other institutions. During the Guelmim floods, better anticipation of a crisis declaration could have allowed the rapid activation of various emergency management agencies using the more limited provincial resources (Zurich, 2015). In confronting more complex crises, many OECD countries have developed forecasting abilities within their crisis units, as has been the case in Denmark, for example (Box 5.11).

Box 5.11. Crisis anticipation in Denmark

The Pandora Unit was implemented by the Danish Emergency Management Agency to support the emergency intervention process by providing advance crisis analysis for crisis managers. In a crisis situation, the Pandora Unit is deployed as part of the crisis management structure. It works independently to provide decision makers with an in-depth analysis of the current crisis and its potential evolution. Pandora's goal is to identify issues that may lead to a deterioration of the crisis, preventing the situation from worsening. These issues may be linked to the response to the crisis, but also to the overall crisis dynamic.

More specifically, the Pandora Unit is made up of three to seven experts with analytical skills and experience in crisis management, often from diverse backgrounds.

First, Pandora provides an outline of the current disaster and the response, and informs decision makers of previous similar events in Denmark or other countries. It provides a general overview of the situation, by developing standard and alternative prospective scenarios, that is, it questions the hypothesis that the crisis will follow a standard trajectory. Finally, Pandora describes the factors that may cause a deterioration of the crisis within a precise timeframe (usually 24 hours).

Source: OECD (2016), Toolkit on Risk Governance, https://www.oecd.org/governance/toolkit-on-risk-governance/ , reviewed 12 February 2016.

The lack of private sector actors in the crisis management process results from the absence of a frame of reference to formalise their participation in emergency response activities. During emergency situations, it is sometimes necessary to call on additional resources beyond those controlled by government agencies. Advance planning that integrates private sector resources in a formal framework will be an indication of improved crisis management efficiency.

Operational communications

While there is no unified crisis communications system in Morocco, the CVC disseminates information among crisis management actors at the central level. Good connections between the CVC and the *wilayas*, with technical networks (meteorological, hydrological and seismic) and their on-the-ground relays helps communicate information and provide "sense-making" or the development of an operational image of the crisis. The Royal Remote Sensing Space Centre's participation in the international space agencies charter aids in this as well, providing access to real-time satellite images of disasters. The activation of an information and communications platform is an essential crisis communications task establishing good communication-sharing among strategic decision makers, reports that facilitate decision making, a chain of command to make sure orders and directives are carried out appropriately and information exchange at the local level among the various emergency response players. Communication systems continuity is another essential issue, since network saturation during crisis periods can result in communications problems. Currently, there are too many parallel and independent reporting processes among the various players, using channels and tools that are not always interoperable.

However, the absence of a unified communication system among actors involved in the response network and the use of different communication channels hinders cooperation on the ground. Coordination between agencies involved in emergency response could be improved by the use of standardised protocols, shared procedures and a common communication tool to establish a crisis dashboard used by all actors in the chain of response. There are too many parallel and independent reporting processes among the various actors, that use channels and tools that are not always interoperable.

Many OECD countries have implemented operational crisis communications standards that can be used by all players within a response network. The CECIS standard is gradually being adopted in Europe. In the United States, the national incident management system is a very effective tool used by all actors of crisis management (Box 5.12). It has improved significantly since Hurricane Katrina, which demonstrated the lack of coordination among the various American agencies in their response to the crisis that affected New Orleans.

Box 5.12. The National Incident Management System in the United States

Since 2005, the National Incident Management System, NIMS, has defined common emergency management protocols for various American agencies. NIMS is composed of a standardised emergency management system implemented at the federal, state, tribal and local levels as well as within NGOs and the private sector, to respond to needs resulting from crises independently of jurisdictional and political borders. With the goal of strengthening inter-agency cooperation and interoperability, NIMS provides standardised procedures for 14 different situations for the direction of intervention, operations, communications, planning, logistics, administrative and financial, intelligence and investigative services. Decision making on goals and action plans is centralised within one command unit to lessen the possibility of contradictory orders and establish sole command responsibility. Organisations can then respond to emergencies in an appropriate and coordinated manner with mutually agreed-upon goals and strategies. Additionally, to ensure effective communications among the players, the system has adopted a common set of terms. Information exchange is facilitated by "public communications officers," who are in constant contact with Command.

Source: OECD (2016), Toolkit on Risk Governance, https://www.oecd.org/governance/toolkit-on-risk-governance/ , reviewed 12 February 2016.

Feedback

Providing post-disaster feedback is a fundamental part of the risk management cycle. The receipt of feedback is essential to improving public policies and continually adapting them to the changing risks and vulnerabilities. As in many OECD countries, significant improvements in crisis and risk management have followed serious disasters in Morocco. Good risk management includes appropriate institutionalisation of feedback and continual improvement in emergency plans and processes based on that feedback.

Many evaluation reports have been put in place by the various emergency response agencies following disasters, but these elements are not collected and used systematically and centrally. ONEE post-crisis evaluations and METS documents published following major events are quite detailed and contain a significant amount of information that is useful in improving the process. The Souss-Mass Draa Wilaya also undertook a detailed analysis of past events to improve its emergency plan. While the CVC has a follow-up tool for disasters and began using evaluation reports in 2008 and 2012, this process would benefit from better institutionalisation to improve procedures.

Inclusion of civil society

To reduce the damage caused by disasters, it is important to involve the public using a targeted communications approach to warn people and keep them informed of efforts underway, measures to take, and the evolution of the crisis. Civil society is also a force that can support the authorities in responding to local needs. Currently, crisis communications and the mobilisation of volunteer organisations are not well developed in Morocco, but there are many opportunities to better involve civil society during a crisis using a societal approach to risk management such as the one proposed in the OECD's recommendation on the governance of critical risks.

Crisis communications directed towards populations

Crisis communication is not properly formalised in Morocco. The CVC, *wilaya* and provincial crisis units usually activate a communications unit to provide statements to media. In light of past crises, it appears their role is often limited to communication about actions taken and provision of an assessment once the crisis has passed. Announcing the end of an emergency is an important point in crisis communications, but using communication methods targeted to the affected population during a crisis should also be a central goal of crisis communications.

Populations can be difficult to reach using traditional communications tools, and the Ministry of the Interior's auxiliary network emphasises direct human communication. Early warning systems also find it challenging to directly reach populations. The network of sirens in place in Civil Defence barracks is rarely used, even for exercises and drills. Morocco has very dense Ministry of the Interior auxiliary networks in many urban neighbourhoods that include pashas and circle chiefs, and headmen and *muqaddams* in rural villages which constitute an essential information tool for the population. These leaders usually relay information related to emergency management, but these information tools are often not used sufficiently to ensure their messages result in actions being taken by the population. During the Guelmim floods local populations refused to leave their homes, located on *oued* beds, because they were afraid of not being able to return or of losing the property they left behind.

A revolution in communications through social media and 24-hour news networks will require Morocco to revisit its crisis communications policy. The use of social media by government agencies to communicate directly with citizens is not yet a reality in the country. However, with 7.2 million users, over a quarter of Moroccans use Facebook, and 77% are aged between 15 and 29 years old (Arab Social Media Report, 2014). Recent crises have shown how extensively Moroccans use social media to relay information about an event. In many OECD countries, emergency services use social media to communicate directly with citizens and several methodologies have been developed for this purpose (Box 5.13). Many useful applications for emergency management were developed directly by citizens during Hurricane Sandy in New York in 2012. These provide precise information about events and needs, detect the actual occurrence of a crisis, and widely inform populations at risk about measures to be taken. Presence on social media is also important to counter the spread of rumours that can sometimes result in undesirable reactions by the population and make crisis management more complex. As an example, rumours of a tsunami were spread in Morocco in 2013, creating panic and requiring a denial to be issued by the authorities (Box 5.14). Developing a relationship of trust with citizens requires on-going use of social media by emergency managers.

Box 5.13. Implementing a social media-based crisis communications strategy

Over the last few years, social media has become one of the preferred means of communication in modern societies, including in crisis situations. In 2012, over 20 million tweets related to Hurricane Sandy were posted on Twitter. Although social media represents a rapid way to communicate, it is also sometimes imprecise, and authorities need to be present and effective on social media during a crisis to counter this. In the United States, the Center for Digital Government has developed seven "best practices" for crisis managers using social media during crisis situations:

1. Identify a consistent hashtag: it is important to give the disaster a name, helping social media users easily find and follow information.
2. Be the voice of authority: it is essential to have a clear and authoritative voice.
3. Establish a rumour control website: since social media facilitates the spread of rumours in a very short time frame, a website for tracking, verifying and correcting rumours is useful.
4. Use several people to manage the flow of requests: because major disasters require continual coverage of social media, it is important to divide the work.
5. Avoid social media scams: there are many social media scammers, and it is important to identify, avoid and blow the whistle on them.
6. Implement a social media archiving solution: this is particularly useful following a crisis, especially for evaluating the actions of the authorities.
7. Choose a precise communications strategy in advance: targets and types of social media must be defined, tasks and responsibilities for communications in the crisis structure must be defined, and accounts listing the organisation's role must be opened.

Source: Center for Digital Government (2014), *Social Media: Emergency Communications' Best Ally*, http://825d0007e19cfb8330f5-793aa0e2839afbbc4a0b9a46376ed589.r13.cf1.rackcdn.com/CDG14_BRIEF_ArchiveSocial_V.pdf

Box 5.14. Spread of tsunami rumours in 2006 and 2013

Over the past ten years, two unfounded rumours of tsunami warnings created panic in Morocco, forcing authorities to deny them.

In May 2006, a statement from the website of the "Centre for Ufological Research", created for the purpose, announced a 200 metre wave approaching the Moroccan coast on 25 May 2006, resulting from comet fragments impacting the Atlantic Ocean. This threat was taken seriously by some people, who passed on the information and planned preventive measures, such as moving inland. As the rumour spread, Moroccan authorities had to deny it: the National Meteorological Office, using an official MAP press release, rejected the claims and ruled out any tsunami danger to Morocco. Additionally, through print media, the Ministry of Communications and government spokesperson also called the rumours unfounded.

In 2013, a second tsunami rumour gained a large following in Morocco. Relayed by the Egyptian site "Al Dostor," based on information supposedly obtained through the High Egyptian Meteorological Authority, this rumour was reported by various media and social networks. This time, a ten-hour-long storm would affect Morocco on 26 December 2013, generating a tsunami. Estimates of human and economic losses were also provided: the forecast tsunami would result in at least 10 million victims and 9 million Dirhams in property losses. This rumour was denied by the High Egyptian Meteorological Authority, which stated it had never communicated such information and Morocco would not be hit by any storms in that time period.

Sources: Actumag.info (2013), A Tsunami Will Hit Morocco on December 26, Actumag. info, 16 December 2013, http://actumag.info/2013/12/16/un-tsunami-frappera-le-maroc-le-26-decembre/, reviewed 18 February 2016. Morocco Today (2006), Tsunami in Morocco: A Giant Hoax, Morocco Today, 23 May 2006, http://www.aujourdhui.ma/une/actualite/tsunami-au-maroc-un-canular-geant-41732#.Vqd-47erSUn, reviewed 18 February 2016.

Involving civil society

When confronted with situations that may significantly impact the country, encouraging organisations such as NGOs to provide rescue assistance, emergency medical aid and support for populations in crisis is an effective way to improve the availability of services. These resources can provide critical support in emergency management as first responders, due to their presence on-site. The involvement of civil society is even more significant if resources available to civil defence are limited, or when there is a prolonged crisis that seriously taxes the response system. In addition to improving emergency response capacity, strengthening these organisations encourages the spread of values such as citizenship and community solidarity, improves the population's resilience, and increases its trust in government agencies.

Despite Morocco's dynamic civil society, involvement in emergency response activities, and particularly the development of volunteer organisations, has remained limited. Very few volunteer organisations participate in emergency management. The Moroccan Red Crescent is the only volunteer aid organisation recognised as a public service for its role as an auxiliary to government civilian and military health services (Moroccan Red Crescent, 2015). The organisation participates jointly with Moroccan Civil Defence during major disasters. During the November 2014 Guelmim floods, the Moroccan Red Crescent mobilised over 600 volunteers on-site for rescue and aid operations, and conducted almost 50 interventions (IFRC, 2014). With approximately 120,000 organisations in 2014 (Ministry of the Interior, 2014), the sector's human potential is significant, and the existing conditions are right to develop new volunteer structures.

The success of recent pilot programmes demonstrates the interest in this approach in remote areas or for large-scale crises, particularly in dealing with specific risks. A project that has been in place since 2008 in collaboration with the Swiss Agency for Development and Cooperation (DDC) and the Directorate-General of Civil Defence mobilised 740 volunteers, composed of 19 groups of local volunteer rescue personnel in the old cities of Fez, Meknes and Sefrou (Box 5.15). During these programmes, there was strong public interest in emergency assistance and participation in emergency management. In the Ourika Valley, local populations took part in developing the flood prevention and warning system. They relayed information to tourists and helped evacuate them to protected areas during flood warnings.

Box 5.15. Local Volunteer Emergency Workers in Morocco

Since 2008, the Swiss Agency for Development and Cooperation (DDC), in partnership with the Directorate-General of Civil Defence (Ministry of the Interior), has led a project to develop a network of local volunteer emergency workers in Morocco. The main objective is to reduce human and economic losses from disasters by building response capacity in both civil society and government agencies, particularly in the most vulnerable old cities. This project addressed three ancient medinas: Fez, Meknes and Sefrou, whose density and narrow streets make emergency evacuations and access for essential emergency equipment difficult. The project was later expanded to Tangier and Casablanca.

Nineteen teams of volunteer emergency workers were formed and 740 individuals (40% of whom were women) were mobilised, trained and equipped. The training provided theoretical knowledge as well as practical skills through the use of on-site simulation exercises. Training sessions covered raising awareness of disaster risks, psychological support, fire prevention, locating victims and rescue techniques. Trained emergency workers received the equipment needed for interventions: uniforms, gloves, hydraulic pumps, fire extinguishers, first aid kits, etc. Additionally, nineteen containers of supplies were placed at the entrances to the medinas to facilitate the work of local volunteer emergency workers. Finally, a programme aimed at raising awareness about risks for the most vulnerable populations was established.

By the end of 2012, local volunteer emergency workers had successfully conducted over 50 emergency interventions. They demonstrated their ability to respond quickly in areas that are difficult to access and reduced the potential number of victims through rapid evacuations. In addition, this improved coordination among the various emergency response players and strengthened the role of civil society. This initiative received widespread public support. The idea is now to expand the project to other parts of Morocco. It is clear an inclusive approach with institutional recognition of the status of emergency workers can be part of a significant improvement in Moroccan emergency response.

Sources: Swiss Confederation / Agency for Development and Cooperation (2014), presentation brochure for the "Local Volunteer Emergency Workers, Phase 4" project, https://www.shareweb.ch/site/Disaster-Resilience/resilience-and-related-topics/Documents/SDC_Secouristes-Volont-Proximite_Maroc.pdf, reviewed 10 December 2015. Emergency Live (2014), Local Volunteer Emergency Workers in Morocco, 7 July 2014, http://www.emergency-live.com/fr/equipements/les-secouristes-volontaires-de-proximite-au-maroc/, reviewed 10 December 2015.

However, the lack of incentives and defined status is an impediment to the development of a volunteer emergency worker system. In other OECD countries, mobilisation of volunteers is based on an established legal framework, connected with an insurance mechanism and compensation for employers. In Italy, this is defined by Presidential Declaration 194/200, which recognises volunteer organisations as an operational component of the civil defence system (Box 5.16). Over 4,000 associations are listed according to their area of intervention (medical services, fire services, telecommunications, etc.) and activities. They are fully integrated in civil defence. In Morocco, a structure for crisis volunteers is needed to align with the best practices identified in the projects and to make them more widespread.

Box 5.16. Mobilising Italian civil defence volunteers

Due to recurring large-scale disasters and the strong solidarity among the Italian public, volunteerism has traditionally been well developed in the country. Dramatic events such as the 1966 floods in Florence and Friuli and the 1980 Irpinia earthquake resulted in the first spontaneous widespread mobilisation of volunteers in Italy. Since then, Presidential Declaration 194/2001 has governed the participation of volunteer organisations in civil defence activities. This declaration stipulates that volunteer associations may be formed by a freely constituted organisation, including municipal civil defence groups, and must be not for profit, democratically structured and established for the purpose of rendering aid. The government contributes to optimising equipment and improving technical preparation and training of volunteers to ensure their safety during emergency operations. In addition, volunteers involved in civil defence activities receive some guarantees: the preservation of their previous government or private sector job, the maintenance of retirement contributions, and insurance coverage. Expenses incurred in connection with emergency operations or training are also reimbursed.

The Italian approach to mobilise volunteers has had good results. Civil defence volunteers are now one of the essential parts of the system: they are a dynamic resource with over a million members across the country and nearly 4,000 organisations on the National Civil Defence list. Volunteer organisations cover the entire country, allowing for rapid mobilisation: 100,000-150,000 volunteers can be mobilised in less than two hours. In fact, volunteer organisations are a very economical way to respond to major disasters. This was demonstrated during the Aquila earthquake, when only 730,000 volunteer workdays were needed for recovery and reconstruction, while using professionals would have cost 100,000 million euros.

Source: OECD (2016), Toolkit on Risk Governance, https://www.oecd.org/governance/toolkit-on-risk-governance/, reviewed 12 February 2016.

Benefits of international cooperation when managing complex crises

International cooperation plays a crucial role in extreme disasters, especially when the national level of preparation and response capacity are overwhelmed by the scope of the damage.

Morocco's growing involvement in the system of international cooperation in civil defence demonstrates the importance of this final ramp-up level for major crises. For example, over the past few years, Moroccan Civil Defence units have been deployed in Algeria, Iran and Turkey (Ministry of the Interior, 2005). They supported local authorities during emergency situations. Likewise, rescue teams from France, Japan, Greece and Austria participated in rescue operations during the Al Hoceima earthquake of 2004. International cooperation in civil defence should increase in the future, since Morocco's civil defence unit has obtained international INSARAG certification in 2014. This certification will be crucial to strengthen the integration of Moroccan civil defence in the international emergency response system. It will allow Morocco to be included in international interventions by civil defence forces organised under the leadership of the United Nations. Certification will also facilitate foreign involvement in major disasters in Morocco.

This support will be even more effective if mechanisms are established for receiving, coordinating and managing international contributions with national efforts already in place. Cooperation under the European Civil Defence Mechanism, as well as with Swiss civil defence and regular joint international exercises are leading in this direction (see below). Without a specific plan of action for international coordination in a major crisis, there is a risk of confusion and reduced effectiveness. Morocco does not have a specific plan of action for international coordination in the event of a major crisis. Consequently, there is a risk of confusion and reduced effectiveness. For example, the Al Hoceima earthquake generated an outpouring of international support. Many countries, as well as the UN and the Red Cross, sent rescue teams, emergency medical resources and survival equipment. However, delays in international aid due to the lack of an international cooperation management mechanism led to frustration among the affected populations. Communities responsible for organising the distribution of aid were often overwhelmed by the amount of aid provided, and could not effectively distribute it to those in need, leading to many more losses.

To respond to these problems, some OECD member countries have implemented mechanisms to coordinate the reception of aid during crises. This is the case in the United States, with the International Assistance System, a Concept of Operations (IAS CONOPS). IAS CONOPS is directed by the U.S. State Department, the Federal Emergency Management Agency (FEMA) and USAID. It established an inter-agency coordination mechanism to handle international aid offers, submit requests, and allocate aid based on needs on the ground (TRIG, 2016). Mexico has implemented a similar system as well as a guide for local players to support cooperation with international NGOs during major disasters. A similar approach could be taken in Morocco to resolve issues that arose during previous disasters.

International crisis cooperation also includes working with the tourism industry by establishing appropriate protocols to cope with tour operators and chancelleries. This includes not only the provision of specific assistance to tourists for their eventual repatriation, but also prevention of new tourists arriving in high-risk areas during crises. Flooding in the southern part of the country in November 2014 demonstrated the need for such a plan.

Conclusion and recommendations

Crisis management capacity has continued to grow in Morocco, especially since 2000, and because disasters occurred and resource inadequacies and delays in interventions made shortcomings apparent. The resources and professionalism of the civil defence has been strengthened. The creation of the Centre for Monitoring and Coordination in 2008 supplemented its resources with a mechanism for decision-making and crisis management in real time.

Government crisis management agencies are preparing to confront a wide range of crises. This system is built on a subsidiarity approach and a centrally coordinated chain of command with decentralised representatives. Although warning, communications and information sharing systems are gradually improving through better coordination, the total resources available for response seem limited for large or very large-scale crises, which are not well anticipated.

An appropriate approach would be to build high-performance emergency response units and countrywide linkages. In practice, civil defence and health services' human and technical resources should be gradually strengthened. This should be done as much as possible to balance the risks, needs and resources. Emergency medicine, hospital contingency plans, bed availability and technical expertise cannot be improvised. A first step may be to target this development in priority areas, such as highly populated urban or suburban areas, or areas that become highly vulnerable at certain times, particularly during the tourist season.

Civil society and local organisations can become more involved in improving the system's ability to deal with the most extreme or unforeseen events, enhancing resources and rapid response capability on-site. In addition to the Red Crescent, Morocco has a significant potential for emergency aid volunteers thanks to its widespread network of organisations. The status of volunteers must be clarified and strengthened to make this a reality. Resources for training and certification will be indispensable. International organisations have already shown readiness to support these efforts.

Qualitatively, the diversity of vulnerable groups will increase over time. Less generic responses must be developed for populations at risk, such as the elderly, ill, handicapped, tourists or immigrants in transit. Early warning systems (understanding messages) and responses require specific preparation (location tracking, adapted resources, informing third parties). Preparation, training and coordination of various actors (transport providers, hotels, telecommunications providers, insurers, diplomatic and cooperation services) are essential.

Recommendations

MOVE INTO OVERDRIVE AND RAISE THE LEVEL OF PREPARATION FOR MAJOR CRISES

Strengthen the Monitoring and Coordination Centre (CVC) by coordinating with warning systems and critical infrastructure.

Activation of the CVC can increase in efficiency as warning systems become more widespread and professional. Similarly, when critical infrastructure operators become part of a network, it will be easier to mobilise them both as potential victims to be protected and as resource centres. To do that, CVC's structure must be strengthened in connection with the current programme of territorial reform.

Pursue the coordination of means and strengthen emergency response capacity in accordance with national risk assessment.

A holistic assessment of all the resources available for emergency response in Morocco should allow comparison with the national risk assessment in order to identify the priority gaps to be bridged. It is possible to initially target priority areas such as those with high-density population areas in urban or suburban areas, or areas that become more vulnerable at certain peak tourist periods, as well as less well-covered risks.

Continue to strengthen the country's early-warning system, specifically for floods and tsunamis.

The meteorological early-warning system must be connected to the development of more precise warning messages for communities and wider dissemination of these messages to interested parties. The coverage of at-risk catchment basins by automated flood warning systems could be a part of this process, as could the completion of a tsunami warning system, associated with education about actions to take in case of a warning.

Develop inter-agency emergency plans for the main worst-case risk scenarios at the national and regional levels.

Review the various emergency plans established in the regions on the basis of identified best practices. Ensure they are regularly updated and tested for inter-agency exercises. Develop in particular an emergency plan for the tsunami risk, for a large earthquake in the north and in Agadir, and for the concomitance of two important floods in the north and south of the country.

Draft a crisis management coordination manual in Morocco specifying the roles of the various stakeholders. The preparation of such a manual will be an opportunity to review the crisis coordination structures to be set up at local level and to clarify crisis leadership under the *wali*.

Develop an active strategy for crisis information and communication using new technologies, particularly social media.

The development of a shared information and communication system with all agencies involved in crisis management is an essential non-structural measure. The use of social media by agencies to communicate with individuals and disseminate local information will significantly strengthen Morocco's crisis communications.

Encourage the development of volunteer groups for crisis management by developing the status of volunteers.

The role of volunteer groups in crisis response must be strengthened in Morocco as an adjunct to government agencies at the local level in the event of a major crisis. Their closer proximity and in-depth knowledge of the area will enhance emergency response. A statute specifying their legal form, involvement and responsibilities along with a training programme will ensure this sector's development.

Strengthen international cooperation activities in emergency response.

International cooperation between emergency response organisations must be pursued to strengthen the coordination abilities of international and regional partners in the event of a crisis. Reflection on methods to receive international support must lead to the establishment of appropriate procedures

Bibliography

Emergency Architects (2004), "Intervention Report, 13 March 2004", Al Hoceima Earthquake, http://www.architectes-fr.com/~contact@archi-ciaf.com/info-ciaf/FOV7-0011C9C5/I00163E6D/__OpenNdx

Morocco Today (2014), Morocco: Associative Explosion, Morocco Today, 29 December 2014, http://www.aujourdhui.ma/maroc/societe/maroc-explosion-associative-115474#.VqoXRU32Z9A reviewed 12 February 2016.

Swiss Confederation / Agency for Development and Cooperation (2014), presentation brochure for the "Local Volunteer Emergency Workers, Phase 4" project, https://www.shareweb.ch/site/Disaster-Resilience/resilience-and-related-topics/Documents/SDC_Secouristes-Volont-Proximite_Maroc.pdf, reviewed 10 December 2015.

Moroccan Red Crescent (2015), Presentation conducted for the OECD Study on Risk Management in Morocco, Rabat, May 2015.

International Federation of Red Cross and Red Crescent Societies (2014), More than 600 volunteers mobilised as heavy floods hit Moroccan cities, kill dozens, 2 December 2014, http://www.ifrc.org/en/news-and-media/news-stories/middle-east-and-north-africa/morocco/more-than-600-volunteers-mobilised-as-heavy-floods-hit-moroccan-cities-kill-dozens-67678/, consulted on 18 February 2016.

Golnaragui M. ed. (2012), Institutional Partnerships in Multi-Hazard Early Warning Systems, Springer-Verlag Berlin Heidelberg.

Khaldi, H. (2005), Emergency Services and Basic Survival Techniques.

The New Tribune (2015), Morocco Participates in a Tsunami Warning Drill, 11 November 2015, http://lnt.ma/le-maroc-participe-a-un-exercice-dalerte-au-tsunami/ reviewed 12 February 2016.

Media24 (2016), 4.5 Billion Dirhams for the Anti-Drought Plan, the Details, Media24, 29 January 2016, http://www.medias24.com/ECONOMIE/ECONOMIE/161383-45-MMDH-pour-le-plan-anti-secheresse-les-details.html reviewed 12 February 2016.

Ministry of Economy and Finance, 2015.

Ministry of National Education, Higher Learning and Management and Scientific Research Training, Management of the Educational Sector after the February 2004 Earthquake.

Minister for Water (2015), Evaluation of the Implementation of the National Flood Protection Plan.

Minister for the Environment (n.d.), Presentation by the National Meteorological Office, https://www.scribd.com/doc/313553983/Dmn-Maroc-Dkhissi.

Ministry of the Interior (2005), Civil Defence Missions, presentation conducted in March 2005.

Ministry of the Interior (2014), Presentation of the New Approach to Crisis Management, Monitoring and Coordination Centre, 20 February 2014, Rabat.

Mourtada R., Salem F., S. Shaer (2014), Arab Social Media Report, 6th Edition, Mohamed Bin Rashad School of Government, June 2014.

Sahara and Sahel Observatory (OSS) (2009), Towards an Early Warning System for Drought in the Maghreb, Synthesis Collection no. 4, http://www.abhatoo.net.ma/content/download/10256/163787/version/1/file/vers_systeme_alerte_precoce_secheresse_maghreb.pdf

OECD (2015), *The Changing Face of Strategic Crisis Management*, OECD Publishing, Paris.

OECD (2016), *Toolkit on Risk Governance*, https://www.oecd.org/governance/toolkit-on-risk-governance/, reviewed 12 February 2016.

National Electricity and Potable Water Office (2014), Management of the Potable Water Crisis in Greater Agadir during the Storms from 27/11/2014 to 30/11/2014, post-evaluation report, no. 113/2014, December 2014.

UNESCO and the Intergovernmental Oceanographic Commission (n.d.), Tsunami Warning System for the Northeast Atlantic and the Mediterranean, presentation brochure.

Souss Massa Draa Regional Wilaya (2014a), "Operational System for Flood Risk Management", *Procedural Framework*, Monitoring and Coordination Unit within the Agadir Ida-Outanane Prefecture, 25 December 2014, Agadir.

Souss Massa Draa Regional Wilaya (2014b), "Storms from 27 November to 1 December 2014", *Monitoring and Coordination System*, Monitoring and Coordination Unit within the Agadir Ida-Outanane Prefecture, 25 December 2014, Agadir.

Souss Massa Draa Regional Wilaya (2014c), "Evaluation Report on the Operational System for Monitoring and Management of Interventions: Storms from 27 November to 1 December 2014", Monitoring and Coordination Unit within the Agadir Ida-Outanane Prefecture, 25 December 2014, Agadir.

Zurich (2015), 2014 Floods in Morocco: What Lessons Can Be Learned from Guelmim and Sidi Ifni?, www.zurich.com/_/media/dbe/corporate/docs/corporate-responsibility/risk-nexus-morocco-floods-of-2014-november-2015.pdf?la=en

Chapter 6.

Recovery, rebuilding and financing

Emergency management by itself is not adequate without recovery policies designed to restore the public's living standards and critical sectors' capacities. Recovery must also be accompanied by reconstruction policies. This chapter covers financial mechanisms implemented by the Moroccan authorities to support recovery, continuity of activities and reconstruction. It includes risk transfer mechanisms, recently established in Morocco, particularly for the agricultural sector. Moroccan authorities set up first initiatives on disaster risk financing, which stresses the importance given to this issue.

Introduction

The short and medium term impacts of disasters are often significant in Morocco. Recent disasters have affected not only individuals, destroyed homes, as was the case during the Al Hoceima earthquake of 2004, but they also destroyed businesses, such as during the 2008 floods in Tangier's industrial district. Disasters have also affected key economic sectors, such as tourism during the Ourika floods in 1995, or agriculture, when 90,000 hectares of arable land were destroyed during the 2009 floods in the Gharb Valley.

Beyond direct damage such as the destruction of homes, damage to business inventories or means of production, disasters can result in medium to long term socio-economic damage if there are no procedures in place for post-disaster recovery and reconstruction. The most vulnerable poor communities at risk from floods, drought or earthquakes may find themselves suddenly without sources of income or capital and forced into rural exodus. Businesses that have not planned for periods of interruption in activities must repair damage to their means of production, rebuild their inventories or remobilise their employees. The longer it takes for recovery and reconstruction, the harder it becomes for an economy to regain its pre-disaster productivity level.

As the OECD has emphasised in the G20 context, governments have a role to play in defining and implementing risk financing mechanisms (OECD, 2012). Damage to or destruction of public infrastructure is usually a significant part of the damage caused by disasters. The indirect effects of disasters on economic activity may also have tax consequences and affect government revenue, especially if economic disruption is prolonged. Once the disaster is over, it is important for decision makers to implement effective, transparent and fair financing mechanisms that promote reconstruction policies, including those for critical infrastructure. They must also support continuity of business and public service activity to accelerate recovery. This represents an opportunity to promote more resilient approaches and reduce long term risk.

While assessing risk management issues, Moroccan authorities have started to consider the implementation of financing strategies for disaster risk management, particularly under the leadership of the Ministry of Economy and Finance. Faced with the economic and financial stakes involved in disasters and losses that may affect different types of players (individuals, networks, businesses, the public sector), there is a need to identify appropriate refinancing or compensation mechanisms for each of these categories.

Financing mechanisms for rebuilding and recovery in Morocco

Morocco's risk management policy has long been based on *ad hoc/ex post facto* approaches for reconstruction and recovery. The establishment of the Fund to Combat the Effects of Natural Disasters in 2009 was a first step in risk financing. Partnership approaches between the government and the insurance industry are the order of the day for establishing risk transfer mechanisms to reduce the economic and social impacts of disasters.

Ex post government compensation, Morocco's principal risk financing mechanism

In Morocco, recovery and reconstruction financing comes primarily from *ad hoc* approaches, combining public financing, domestic donations and international donations. While recovery and reconstruction are mostly the government's responsibility through a combination of compensation methods, other resources are also used. After the Al Hoceima earthquake, 3bn Dirhams were mobilised for recovery and reconstruction efforts in affected areas: 1.39bn from government budgets, 867 million from domestic donations and 755 million from international donations (Figure 6.1).

Figure 6.1. Recovery and reconstruction financing sources for the Al Hoceima Earthquake (in millions of Dirhams)

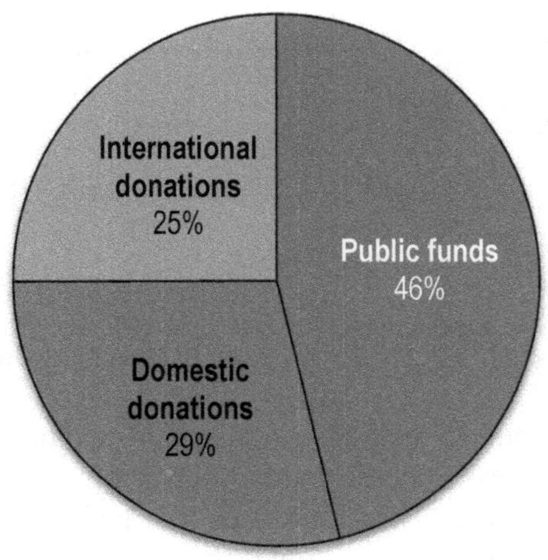

Source: Sehimi, M. (2014), *Construction Sites for a Billion*, Maroc Hebdo, 8 May 2014, http://www.maghress.com/fr/marochebdo/118431, reviewed 19 February 2016.

Several agencies are responsible for the government's recovery and rebuilding efforts. Ministries assigned to specific sectors make budgetary reallocations and mobilise their own resources. Following the Al Hoceima earthquake, approximately ten public agencies made investments in reconstruction (Figure 6.2). The Ministry of National Education allocated 83 million Dirhams to rehabilitate 89 schools and rebuild 15 others. The Ministry of Health participated in the rehabilitation and rebuilding of hospitals in the Al Hoceima Region, investing 81 million Dirhams. The Ministry of Habous and Islamic Affairs helped rehabilitate 45 mosques and rebuild 11 others, with an overall investment of 34 million Dirhams.

Figure 6.2. Public funding for recovery and reconstruction following the Al Hoceima Earthquake by sector

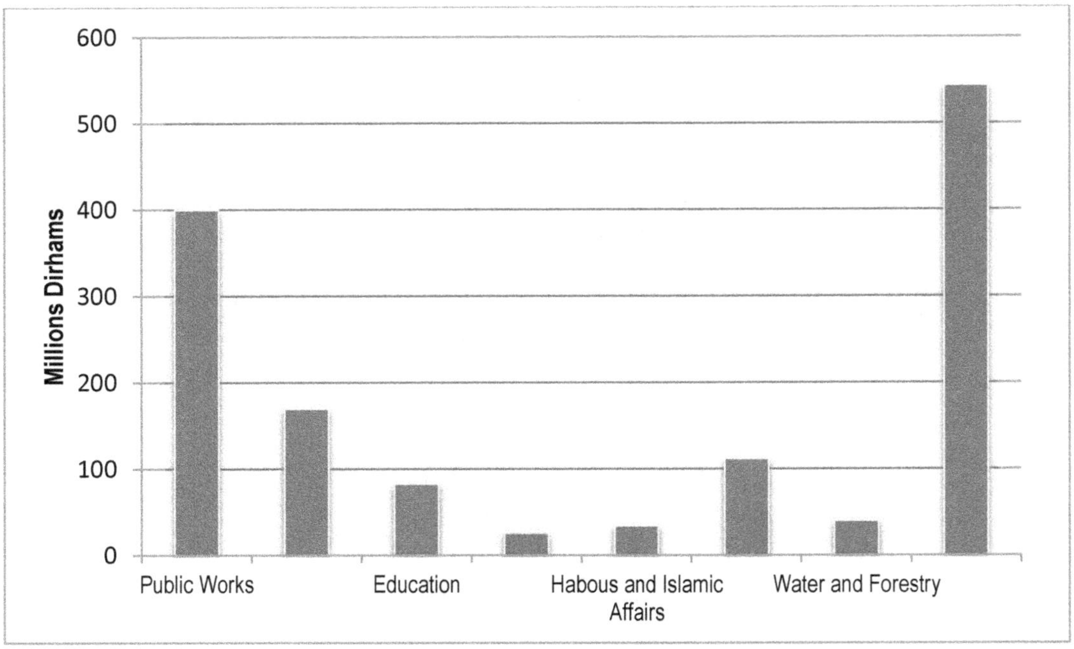

Source: Sehimi, M. (2014), *Construction Sites for a Billion,* Maroc Hebdo, 8 May 2014, http://www.maghress.com/fr/marochebdo/118431, reviewed 19 February 2016.

However, it is important to emphasise that this sector-based resource allocation results from the *ad hoc* process. There is no approach for coordinating and streamlining the various public agencies' activities. This leads to a lack of transparency and clarity in funding allocation. The government has attempted to resolve this by creating the Fund to Combat the Effects of Natural Disasters.

Creation of the FLCN: towards more streamlining of available resources

To standardise and provide better coordination of post-disaster recovery and reconstruction expenditure, a special fund, the Fund to Combat the Effects of Natural Disasters (FLCN), was established in Morocco in 2009.

The FLCN is a special appropriations account managed by the Ministry of the Interior. It was established by Article 16(a) of Finance Law 40-08 for fiscal year 2009 and was supplemented by Article 21 of Finance Law 115-12 for fiscal year 2013. The FLCN allows the coordination of the financing of Moroccan risk management policies. Since its establishment, FLCN has been funded by the government budget at 200 million Dirhams per year (Table 6.1).

Table 6.1. FLCN financing (2009-2016)

Employment programme for 2009-2012	
Gifts from Saudi Arabia	861 MDH
Hassan II Fund for Economic and Social Development	280 MDH
General Government Budget	800 MDH (200 MDH annually between 2009 and 2012)
Financing for the period 2013-2016	
General Government Budget	800 MDH (200 MDH annually between 2013 and 2016)
Hassan II Fund for Economic and Social Development	20 MDH (in 2013)

Source: Ministry of Economy and Finance, Report on Special Treasury Accounts, Ministry of Economy and Finance, 2012, 2013, 2014, 2015.

As the primary budget mechanism dedicated to *ex ante* risks, the FLCN is a first step in a disaster risk financing strategy in Morocco. Its budget gives the government access to financial resources for the improvement of interventions by the various departments, aid and support to affected individuals, recovery and reconstruction operations for damaged infrastructure, as well as strengthening warning and monitoring systems and the construction of prevention facilities. The FLCN has a broad and flexible mandate, contributing to priorities for prevention, emergency response, recovery and reconstruction. Since 2012, significant balance has been carried over each year in the FLCN account, which demonstrates that a substantial part of the resources allocated was not spent (Figure 6.3). Reports on special treasury accounts also show a gap between planned and actual expenditure: for example, while 210 million Dirhams were allocated to improve digital forecasting models and develop meteorological warning systems, only 135.5 million were actually used.

Figure 6.3. FLCN expenses and resources from 2009 to 2014

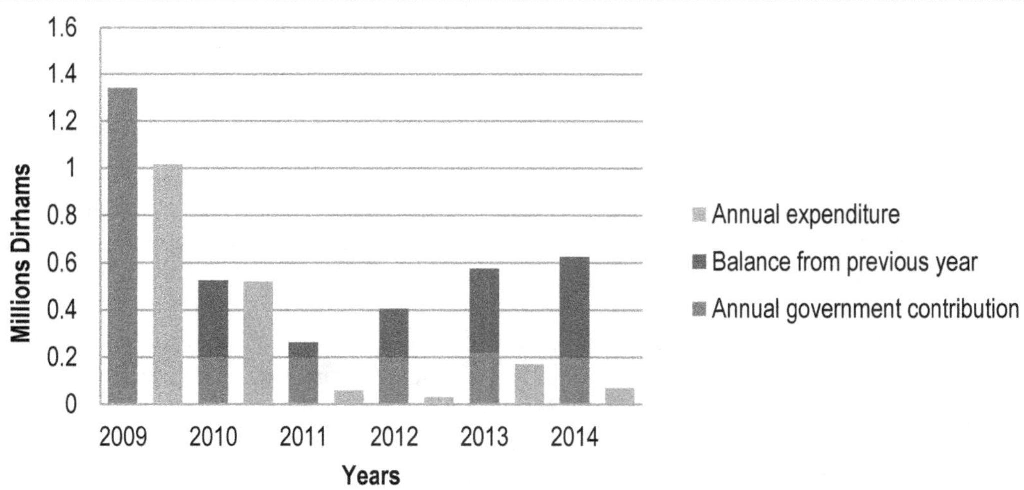

Source: Ministry of Economy and Finance, Report on Special Treasury Accounts, Ministry of Economy and Finance, 2012, 2013, 2014, 2015

More specifically, recovery and reconstruction expenditure represented 42.8% of the Fund's distributions between 2009 and 2013, divided between public infrastructure, agriculture, housing and support of the private sector (Figure 6.4). The Fund participated in recovery efforts for the public, businesses and the government. It contributed to financing repairs for natural disaster damage to roads, trails, civil engineering structures, as well as irrigation infrastructure. Through funding from the Central Guarantee Fund, FLCN also helped implement an emergency housing programme for flood victims and provided funds for affected businesses. The Fund has since financed an equal proportion of *ex ante* prevention disaster risk policies and *ex post* reconstruction and compensation policies. More recently, FLCN has contributed both to *ex ante* risk prevention policies and *ex post* reconstruction policies. However, in the future, it seems riority should be given to financing prevention activities through a dedicated request for projects and by adopting a large-scale investment approach such as the Casablanca west super collector project.

Figure 6.4. FLCN funding distribution (2009-2013)

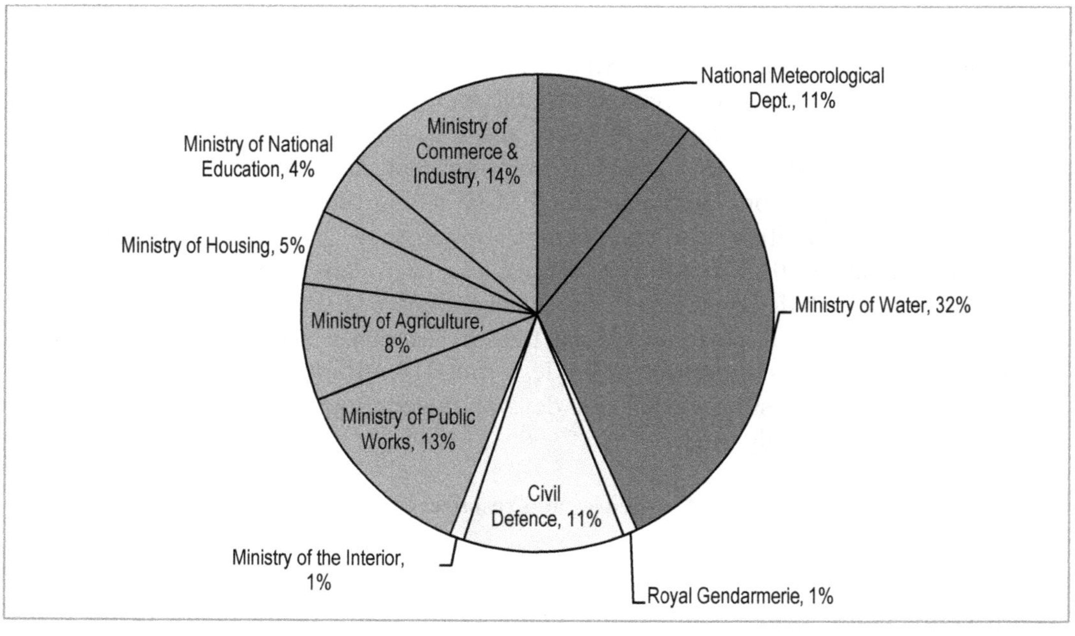

Legend: Light blue: increased intervention resources; medium blue: recovery and reconstruction; dark blue: prevention, monitoring and warning systems.

Source: Michel-Kerjan, Erwann O. (2014), CAS-FLCN: Information items for the definition of priorities and selection criteria for projects submitted for financing and operational implementation, report produced as part of the World Bank's assistance to the Ministry of the Interior.

Nevertheless, with approximately 200 million Dirhams of annual funding, the amounts provided by the FLCN seem relatively low compared to the average annual damage caused by disasters. The World Bank estimated that annual disasters could cause government-related public infrastructure damage of 850 million Dirhams for earthquakes, 4,177 million for floods, and 124 million for tsunamis.

The adoption of an integrated natural disaster risk and resilience programme in April 2016 that consists of a results-based loan programme for 200 million dollars from the World Bank could help strengthen the FLCN's activities (Box 6.1).

Box 6.1. The World Bank provides a 200 million dollar loan to the government of Morocco to strengthen integrated risk management

On 20 April 2016, the Board of Directors of the World Bank approved a 200 million dollar loan to the Moroccan government to finance the "integrated natural disaster risk and resilience management programme." Covering a five-year period, this programme aims to improve the institutional structure for financing disaster risk reduction activities and the financial resilience of populations affected by disasters. The programme has three purposes:

1. Promoting institutional reform and capacity building for risk management, in particular by redirecting the Fund to Combat the Effects of Natural Disasters towards financing prevention projects;
2. Encouraging the development of disaster risk reduction activities, particularly by financing eligible natural disaster risk reduction projects;
3. Improving disaster risk financing and insurance, particularly through a legislative and regulatory insurance structure for catastrophic events and the implementation of an Aid Fund for such types of events.

The World Bank's financing instrument for this project is a "results-based loan programme (PPR)", whereby funds are disbursed based on results achieved.

Source: Ministry of General Affairs and Governance (2016), Note on the integrated natural disaster risk and resilience management programme.

In parallel with FLCN, additional financial resources for natural disasters are mobilised by the government on an *ad hoc* basis for public infrastructure reconstruction, to compensate for individual losses, or to support business recovery. The scope of this funding is, however, not clear. Public infrastructure is often rebuilt using construction funds from the sector Ministries or their agencies. This was the case for METL, which has financed more than 490 projects for a total of 3,206 million Dirhams since 2008 to repair flood damage to roads (METL, 2015).

Similarly, compensation mechanisms for households often include housing solutions using funds from several social programmes, such as the Fund for Housing Aid and Urban Integration (FSHIU) or Al Omrane Agency programmes. Agriculture sector compensation is managed by the Ministry of Agriculture, which can provide a range of direct and indirect grants to farmers to compensate for losses from various kinds of risks. Coverage of customs duties to import spare parts and basic equipment necessary for restarting activities of affected businesses was made available by the Central Guarantee Fund, and it was partly financed by FLCN. Other one-time resources may also be used (e.g. royal gifts, individual gifts, international emergency aid).

Towards risk transfer mechanisms

The use of insurance mechanisms or other financial risk transfer tools is still limited in Morocco. Consequently, damages resulting from major disasters often become the government's responsibility and weigh heavily on its budget. The insurance coverage rate remains very low, around 3% for individuals and 15% for businesses. Additionally, traditional insurance does not cover disasters. The sector's penetration rate (premiums as a percentage of GDP) reached 3.14% in 2014, a rate similar to that of other developing countries but well below that of OECD countries. As a consequence, existing risk transfer mechanisms are currently based on aid from the insured to the uninsured.

However, Morocco's insurance market is currently growing at a rate of nearly 6% per year (premiums increased by 6.08% in 2014, for example). In parallel, the needs of individuals, businesses, and sometimes of large construction projects, are also growing as the economy develops.

On this basis, the Ministry of Economy and Finance (Insurance and Social Contingencies) and the SCR insurance company have collaborated in developing draft Law 110-14, making disaster coverage, such as for earthquakes, tsunamis or floods, required in certain private insurance contracts with an additional premium (Box 6.2). This draft law also establishes the "Aid Fund for Catastrophic Events" which will benefit individuals without coverage. The Fund will be financed by revenue from a special tax, established by regulation, and would receive an initial endowment from the "Insurance Aid Fund" (collecting 10% of the profits from Moroccan insurance companies and providing a response mechanism for potential specific deficits in a required insurance branch).

The draft law establishes a system for covering the consequences of disasters to improve compensation and the recovery and reconstruction process. It must be accompanied by a damage assessment procedure for insurance companies, since their access to affected areas for conducting assessments is not yet systematic. In addition, this legal procedure should be complemented by incentives to reduce risk and temper the moral hazard due to the fact that insurance premiums financing this system don't depend on the risk level. To resolve this issue, using MnhPRA, a risk assessment tool implemented across the kingdom and covering major risks, will be particularly important.

Box 6.2. Draft Law establishing a coverage system for catastrophic events

The use of insurance mechanisms for risk transfer remains limited in Morocco. Traditional insurance does not include coverage of disaster risks, making compensation and the recovery and reconstruction processes difficult.

Currently, the expansion of CATNAT insurance is encountering several problems. The insurance coverage rate is still very low, at around 3% for individuals and 15% for the commercial sector, and traditional insurance does not cover disaster risks. Furthermore, policies are still quite limited and expensive for individuals. Consequently, only large corporations can benefit from this insurance and must pay significant additional premiums.

To implement a more efficient compensation system and reduce the government's budget, the Insurance and Social Contingencies Department of the Ministry of Economy and Finance has taken part in preparing a draft law establishing a coverage system for the consequences of disasters. This draft law envisions requiring coverage of disaster risks in private insurance policies, thereby guaranteeing a minimum right to compensation for bodily harm or loss of the use of a principal residence in a disaster. Through the establishment of the "Aid Fund for Catastrophic Events", this system is based on a mechanism for compensation of disaster victims that combines an insurance system for individuals with an insurance policy and an allotment system for individuals with no coverage.

The Governing Council adopted this draft law on 4 March 2016. Parliament is set to approve it in 2016, with implementation set for the following year.

Box 6.2. Draft Law establishing a system of coverage of the results of catastrophic events *(continued)*

Source: Ministry of Economy and Finance (2015), Note on the presentation of the draft law establishing the coverage system for the consequences of disasters and modifying and supplementing Law no. 17-99 on the insurance code. http://www.sgg.gov.ma/portals/0/AvantProjet/118/Avp_loi_cat-nat_Fr.pdf, reviewed 17 December 2015.

One exception to natural disaster risk transfer for the agricultural sector in Morocco should be noted. In this specific case, compensation and continuity support mechanisms are well-established, with both targeted programmes and specifically developed and innovative insurance products that are widely subsidised under the Green Morocco Plan (*Plan Maroc Vert*). The Ministry of Agriculture and Maritime Fishing (MAPM) has chosen an anticipatory approach. It uses risk prevention programmes as well as new insurance products specificly aimed at reducing farmers' vulnerability to agricultural risk, improving agricultural practices and increasing yields, increasing revenue for farming populations, promoting investment in agriculture and improving the business climate in agriculture. A study mandated under the Green Morocco Plan has identified and ranked the risks to which various agricultural sectors are subject. Drought was identified as the most significant risk, ahead of diseases and pests and high temperatures (Figure 6.5). The study also foresaw the implementation of a wide-ranging insurance system that would cover the principal agricultural sectors throughout the country.

Figure 6.5. Agricultural risk map

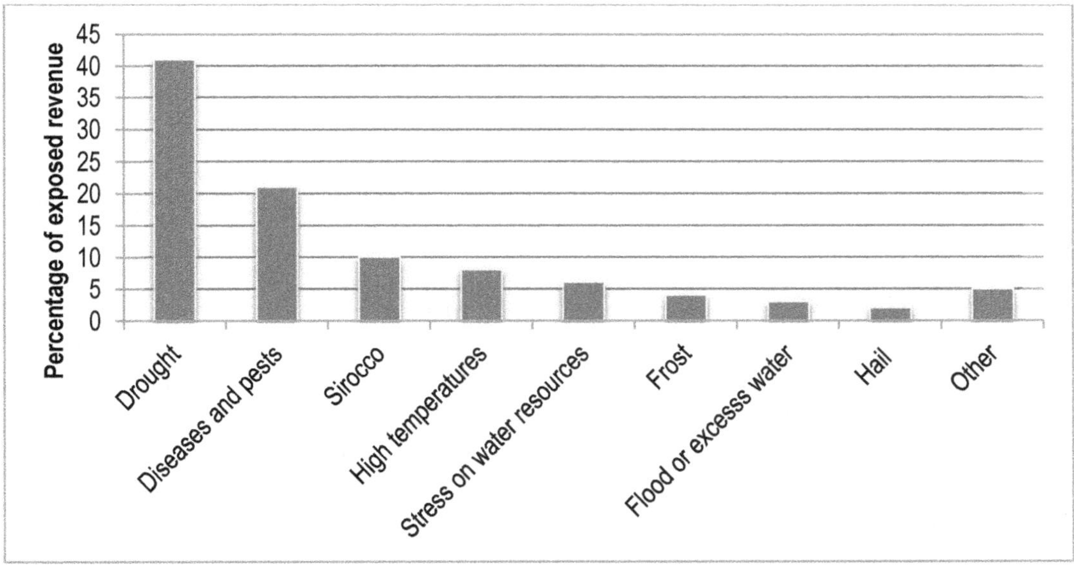

Source: El Yacoubi, M. (2014), Risk Management Strategy in the Agricultural Sector, presentation made to the International Conference on Insurance and Reinsurance of Agricultural Risk, 30 January 2014, Marrakesh.

Based on these findings, multi-risk climate insurance, marketed by the MAMDA insurance company, provides comprehensive coverage to farmers. It is intended for two principal agricultural sectors in Morocco: grain growers (soft wheat, hard wheat, maize and barley) and vegetable growers (broad beans, peas, lentils, chickpeas and haricot beans) and, since the 2013/2014 season, also covers orchard fruit growers. It covers the following climatic risks: drought, excess water, hail, frost, high winds, and sand storms. An extension is planned for orchard fruit growers for the risk of hail, frost, high winds, sirocco (hot winds), excessive heat and excess water.

The success of multi-risk climate insurance is growing, particularly for small farmers, who are the most affected by climate hazards. In 2015, it covered 717,000 hectares. Its short term goal was to ensure coverage of 1 million hectares, representing one-third of Morocco's agricultural surface area (Figure 6.6). Its development was made possible by a 2011 agreement between the Ministry of Economy and Finance, the Ministry of Agriculture and Maritime Fishing and MAMDA. This agreement allowed the government to subsidise between 53% and 90% of this multi-risk climate insurance, based on the areas involved and the coverage level.

Figure 6.6. Surface area covered by multi-risk climate insurance

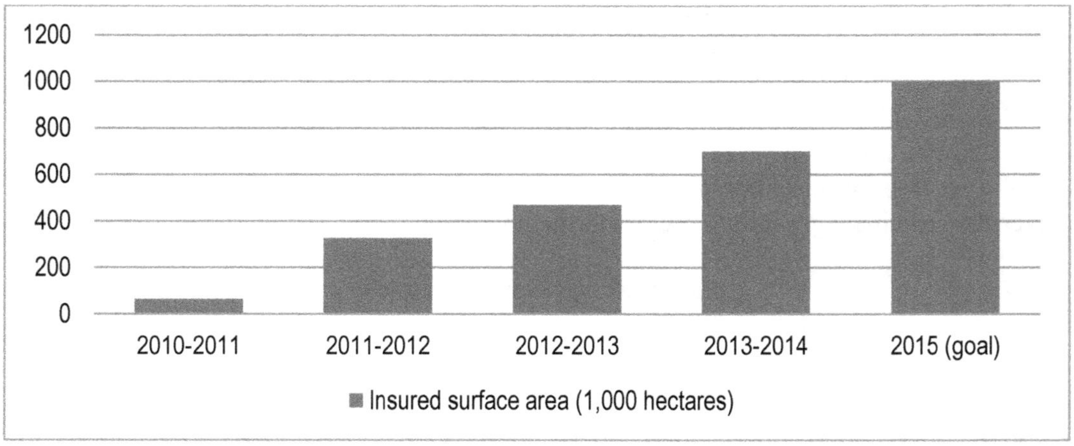

Source: El Yacoubi, M. (2014), Risk Management Strategy in the Agricultural Sector, presentation made to the International Conference on Insurance and Reinsurance of Agricultural Risk, 30 January 2014, Marrakesh.

At the same time, a pilot project was launched to test the feasibility of parametric insurance as a supplement to multi-risk climate insurance. Covering grain production and only for the risk of drought, this pilot project aimed at developing an insurance product in which automatic compensation would be made based on agro-climatic criteria. This pilot project was put in place in three agricultural regions, Chaiouia-Ouardigha, Meknes-Tafilalet and Fez-Boulemane. It was financed by the French Fund for the Global Environment (FFEM) and the French Development Agency for a period of three years, and beginning during the 2013/2014 agricultural season. Once completed, the pilot project should confirm the level of interest in parametric insurance in Morocco and complete the development of a permanent index method to evaluate yields on the community level. To accomplish this, the INRA, MAPM and the National Meteorological Office are contributing to the development of the "Moroccan System of Vegetation Monitoring and Harvest Forecasting". It is a monitoring system for crop development and yield forecasting that uses agro-meteorological data that the parametric insurance product will be based on.

The OECD/G20 risk financing structure

Thus, a structure for risk financing is gradually being established in Morocco, with the implementation of a compensation system for individuals and businesses using a risk transfer mechanism (still to be created), as well as dedicated mechanisms for agriculture.

However, this structure seems to lack both a component for government resources, which could potentially lead to the ability to mobilise significant budget resources for reconstruction following major disasters, and a specific plan for public infrastructure, which is not currently insured. The results of the MnhPRA for public infrastructure damage assessment will be interesting, because they will develop a basis for evaluating risk coverage mechanisms. However, the recurrence of large-scale events over the long term, that are likely increased by the effects of climate change, may question the approach's durability.

In 2012, the OECD and the G20 developed a method for assessing and financing disaster risks (OECD, 2012). It revealed strong ties between risk assessment, mitigation and associated financing strategies, which are essential elements for dynamic and effective risk management strategies. In particular, this methodology demonstrated three important points for risk financing:

- It is essential to precisely measure exposure to risk and the capacity to assume that risk prior to developing associated financing strategies. Based on risk assessment (Chapter 3), such an analysis will identify the economy's financing vulnerabilities and the need for financing tools.
- The availability, appropriateness and effectiveness of financing and risk transfer tools must be evaluated for financial soundness and maturity of the insurance market in a given country.
- It is important to determine if government intervention in financing and risk transfer is appropriate. If so, it is also essential to fix the amount of risk to be shared. Consequently, if government intervention is deemed necessary, the authorities must develop a plan based on the private sector's ability to assume risk, the acceptance of potential negative impacts of the intervention, and the inclusion of appropriate risk sharing mechanisms.

Effectiveness of the rebuilding and recovery process

Whatever risk financing mechanisms are established in Morocco in the coming years, it appears the government will continue to play a front-line role in these mechanisms, if only as guarantor of the continuity between emergency management, recovery and reconstruction of public infrastructure. Beyond the question of mobilising resources, effective public policies and mechanisms are absolutely necessary to accompany and facilitate recovery and reconstruction. These policies must also ensure a return to normal life and economic activity as quickly as possible. More specifically, this includes the question of systems for damage assessment, the need to provide the best level of transparency in compensation mechanisms for individuals and for businesses, the creation of systems of accompaniment dedicated to restarting business activity, and integrating issues of resilience into the reconstruction process following disasters.

Post-disaster damage assessment

In Morocco, damage assessment is conducted by sector commissions established at the provincial or *wilaya* level. They are then sent to the central level. There is no national system that centralises the information gathered, which hinders a comprehensive damage assessment. During significant events, such as the Al Hoceima earthquake, multidisciplinary teams of experts were sent on-site to perform a detailed and coordinated damage assessment. However, this approach is not systematised during smaller-scale disasters such as recurrent flooding.

The process of post-disaster damage assessment would be simplified by the implementation of mechanisms to systematise and facilitate access by private insurers to affected areas. Private insurers have a network of experts to identify and quantify property damage, and repair methods. In Morocco, expert access to disaster areas is not systematic and can be difficult. Implementing a coordinated approach with stakeholders, such as the Ministry of the Interior, could improve post-disaster damage assessment.

Transparency and fairness in resource allocation

Ensuring transparency and effectiveness in compensation mechanisms is also essential. Mobilisation of specific agencies and *ad hoc* government resources is often an appropriate means of managing funds and reconstruction programmes. The enormous reconstruction programme for the city of Al Hoceima after the 2004 earthquake was managed by the Northern Provinces Development Agency. It was able to best ensure appropriate management of funds and project implementation. In these types of situations, the significant monetary amounts involved and the high expectations of the affected population require finding a balance between effective and rapid execution and meeting specific local needs. During the November 2014 floods, monitoring commissions for reconstruction established in the Agadir Wilaya and Haouz province ensured a detailed evaluation of needs and of all stakeholders' respective contributions.

Nevertheless, compensation and reconstruction mechanisms currently in place suffer from a lack of clarity regarding precise rules for accessing funds. Damage assessments are conducted by commissions at the provincial or *wilaya* level and are usually sent to the central level. The allocation of financial resources seems discretionary, which leads to a loss of effectiveness. The chosen mechanism is based on *ad hoc* processes following each crisis between individual damage compensation, reconstruction of public works and support for restarting economic activity. In addition, there is no accounting mechanism for domestic or international donations. Historically, this has led to public discontent, such as during the Al Hoceima earthquake recovery.

Reconstruction and resilience

The government is therefore the principal actor in the reconstruction process and is responsible for ensuring the reconstruction process is an opportunity to reduce vulnerability to future events. This approach includes taking into account short term reconstruction needs and investments for improved resilience over the long term.

The Moroccan authorities have sometimes taken measures to improve resilience in reaction to damage caused by disasters. Some remarkable initiatives to help avoid repeating the same mistakes include the establishment of "no-build" zones or rebuilding more resilient infrastructure. For example, the reconstruction of the city of Agadir, which was 75% destroyed in the 1960 earthquake was done according to the first seismic standards established in Morocco. Some areas in which the destruction was clearly linked to soil instability were declared unsuitable for building and have remained so. Following the Al Hoceima earthquake the city's reconstruction was an opportunity to improve its resilience by using RPS paraseismic regulations. The reconstruction of railways damaged during the Gharb flooding in 2009 will result in its protection being raised to the 100-year flood level. The National Railways Authority (ONCF) implemented an annual budget for investments in eliminating or mitigating identified natural risks. Following flooding, some *douars* were rebuilt outside flood zones, supplementing preventive construction, such as following the Mghogha floods in Tangier in 2008 and 2009 (Box 6.3).

However, compensation mechanisms for damage to existing structures caused by disasters do not include formal and systematic incentives to improve resilience. The initiatives cited above should be encouraged by specific programmes or labels that highlight their purpose. Multi-actor management mechanisms will facilitate the implementation of reconstruction measures and guarantee more efficient use of public funds, including through greater transparency in the allocation of expenditure.

Box 6.3. Resilience of Morocco's industrial districts

The resilience of Morocco's industrial districts is a critical issue. Over the last few years, Morocco has completed several projects to strengthen and make its industry more competitive, particularly with "Emergency I and II" sector plans (2005-2015) and the Industrial Acceleration Plan (2014-2020). However, recent flooding has revealed the vulnerability of strategic industrial districts. It has also shown questions of risk management and resilience have not been sufficiently accounted for.

For example, the Mghogha (Tangier) industrial district, built in 1975 on Oued Mghogha bed, was hit by significant flooding in 2008 and 2009, negatively affecting the activity of the 1,000 businesses located there. Similarly, the Sidi Bernoussi (Casablanca) industrial district experienced major flooding in 2010. According to Izdihar (the Sidi Bernoussi Industrial District Economic Operators Association), 80 of the 600 businesses located at the site were declared to be severely affected.

Following these catastrophic events, measures were taken to improve resilience. In Mghogha, an in-depth risk analysis was conducted by the Water Ministry, and follow-up measures were decided upon. Two dam construction projects and a project to enlarge Oued Mghogha's channel, which would make it possible to control the *oued's* flow rate and eventually its course, were chosen. However, the progress of these projects has been slowed due to lack of financing. In Sidi Bernoussi, the authorities have made several requests for proposals for constructing a new sanitation system, and creating a sewer system to meet the district's needs.

The issue of resilience in Morocco's industrial districts should be placed at the centre of government strategies to guarantee the durability of these new instruments of growth and development.

Sources: Interviews conducted by the OECD for the Study on Risk Management in Morocco (May 2015). Tangier Industrial District Association (2015), website of the Tangier Industrial District Association, http://www.azit.ma/index.php?r=qui-sommes-nous&p=plan-de-la-zone-industrielle, reviewed 14 December 2015. El Azizi, A. (2010), *Tangier – The Black Book of Floods, Issue 31*, 23 January 2010, http://www.actuel.ma/index.php?option=com_magazines&view=detail&id=77, reviewed 14 December 2015. *Behind the Scenes in Tangier, "Tangier: Flood Project"*, http://www.lescoulissesdetanger.com/actualites/124-tanger-se-mobilise-contre-les-inondations-pour-mettre-a-labri-son-tissu-industriel. Sidi Bernoussi Industrial District Economic Operators Association (IZDIHAR) (2010), "The Sidi Bernoussi Industrial District Bandages its Wounds", 12 May 2010, IZDIHAR, http://izdihar.info/uploads/presse/f53.pdf.

Continuity in business activity

Continuity of private sector activities is a major issue in the recovery and reconstruction process. Establishing activity continuity plans can reduce disruptions that may be caused by disasters by maintaining the flow of goods and services. This is even more important in the case of critical infrastructure, such as transport, water, electricity and public services, which are at the heart of a modern functioning economy. Following a disaster, the recovery of a region or a country depends greatly on the ability of its businesses to maintain their level of activity, which can be strengthened by the implementation of specific accompanying government programmes.

It is important to note Morocco currently has no on-going formal public policy system to support business continuity. Although support mechanisms could have been implemented in a timely manner, for example by providing guarantee funds, businesses have been able to limit or compensate for losses resulting from disasters and restart their activities more quickly via the development of voluntary continuity measures. Additionally, there is no set of incentives and no information sharing and exchange forum for businesses, which hinders such a comprehensive approach.

However, faced with the economic and financial stakes involved in disasters, Morocco's large businesses have increasingly become voluntarily involved in the issue of business continuity. This involves evaluating the risks to which their businesses are subject, modelling potential impacts and planning for continuity. For example, multinational corporations with an incentive to apply international standards, such as ISO 31000 (risk management) and ISO 22301 (continuity of activity), have access to guarantees from their international business partners. Similarly, the *Office Chérifien des Phosphates* (OCP) has established a risk management department, responsible for identifying business vulnerabilities, analysing risks, evaluating action plans for potential disasters and developing a business continuity plan. The banking sector, following a directive on continuity of activities from the credit arm Morocco's central bank, regularly develops and submits activity continuity plans to Bank Al Maghrib based on a defined set of rules.

This new awareness of business continuity can also be observed among critical infrastructure operators, both public and private. Critical infrastructure resilience (e.g. energy, transport, communications and information systems) is an increasingly important issue: in a globalised and interdependent world, these systems are indispensable to the proper functioning of the economy. In the energy sector, the delegated corporations of Lydec and Redal, have agreed to ensure their services in a sustainable way and to adhere to the highest standards for security and environmental protection. Since then they have been implementing the necessary measures to achieve this goal. ONEE has also developed a business continuity plan. Concerning the transport infrastructure, the Ministry of Public Works, Transport and Logistics, through the National Airports Authority, the National Ports Authority, the National Railways Authority and the Roads Department, has implemented plans for continuity of activity and an emergency management system to evaluate damage and start reconstruction as early as possible. Finally, in the area of telecommunications, Maroc Telecom has a "Continuity of Service Procedure" to help ensure continuity of service in the event of a disaster.

Although large corporations have deployed efforts to develop and implement continuity plans for their businesses, these appear uneven when considering small and medium-sized businesses (SMEs). Indeed, the continuity of activity is a critical issue for

SMEs. In Morocco, SMEs represent over 90% of the industrial base and 40% of industrial revenue. They remain very vulnerable to the risk of natural disasters, as demonstrated by the 2008 floods in Tangier, which severely affected businesses in the Mghogha industrial district. Following that event, the government asked the Central Guarantee Fund (CCG) to contribute to recovery of businesses following disasters for the first time by entrusting the management of a guarantee fund of 200 million Dirhams from the Fund to Combat the Effects of Natural Disasters (FLCN). One quarter of these funds, or 50 million Dirhams, was transferred to the National Agency for the Promotion of Small and Medium-sized Businesses (ANPME). The agency addressed affected businesses in the Mhgogha industrial district in Tangier by covering customs duties for importing spare parts and basic materials required to restart activities, to the amount of 25 million Dirhams. For the most vulnerable SMEs, the CCG also implemented debt rescheduling and reimbursement of 24 months of debt. However, these programmes are only used occasionally. In Morocco, as for many OECD countries, a real consideration of continuity of activity occurs mostly in large corporations rather than in SMEs. Innovative programmes have been implemented to support small businesses in their rebuilding efforts, as was the case in the United States following Hurricane Sandy (Box 6.4).

Box 6.4. A single clearinghouse for business recovery after Sandy in New York

After Hurricane Sandy in October 2012, the American authorities deployed a network of "Business Recovery Centres" to provide a single clearing house for resources and services from local, federal and central agencies following natural disasters. These centres brought together representatives of the Federal Emergency Management Agency (FEMA) and the U.S. Small Business Administration to answer questions and assist businesses whose activities were affected by the hurricane.

The centres provided a single location where business owners could find a wide array of services:

- Find information about financial assistance offered by the authorities following the disaster.
- Receive help filling out loan request forms.
- Receive expert advice to re-create lost financial records in order to document losses.
- Locate replacement facilities.
- Obtain information on clean-up options.
- Develop an emergency response plan and a business continuity plan for responding to potential future disasters.
- Obtain assistance in strategic planning for rebuilding the business.

Source: New York State Small Business Development Center (2016), SBDC Services for Small Business Affected by Disaster, SBDC website, http://www.nyssbdc.org/disaster.html, reviewed 20 February 2016.

Conclusion and recommendations

Recovery and reconstruction operations are decisive components of an integrated risk management strategy. A large part of the mitigation of economic, social and even psychological consequences for individuals and businesses depends on the speed of service restoration. Medium and long-term traumatic consequences for individuals, families and business owners will be reduced based on the response's speed and adaptation. However, speed must not be a synonym for improvisation or lack of quality.

Since important decisions must be taken in real time, these must be adequately anticipated. When planning is insufficient, there is a danger victims (individuals, businesses, communities) may rebuild on exactly the same site in a flood or seismic zone. This was recognised in Morocco since the 1960 earthquake in Agadir. It is also one of the major issues that led the OECD (Recommendation 5.4) and the United Nations to promote the "build back better" concept (Priority 4, Third World Conference on Disaster Risk Reduction in Sendai 2015). It also represents a potentially "positive" aspect of destruction/reconstruction, which can enable better, more sustainable construction that is less exposed than existing buildings.

Beyond the drive for "better", questions should be raised regarding financing service and functionality restoration, and reconstruction. From a reactive response (general budget, request for international aid), a country can move to the creation of dedicated mechanisms, then to insurance solutions with or without a market, and progressively include diversification in risk transfer (captive insurance companies, catastrophic coverage) while requiring stakeholders to share the burden.

Over recent years, mechanisms for reconstruction and recovery have been under development in Morocco. Since 2009, the establishment of the Fund to Combat the Effects of Natural Disasters has provided the government with a financing structure dedicated to risk management policy, including recovery and reconstruction. This fund, in addition to its effectiveness on the ground and its wide range of interventions, will be a genuine learning tool to allocate resources between projects and different sites.

Improved streamlining of financial instruments available for reconstruction or compensation is also being implemented, along with the development of insurance mechanisms. Heavily subsidised by the government, in the agricultural sector, these insurance mechanisms have provided Morocco with an almost unique experience in building tools that combine the principles of mutual insurance for actors able to access insurance and the principle of aid for those who are not.

Such a combination of insurance principles and aid will be extended (Law 110-14) to individuals. This law also envisions the extension of the range of required insurance. As in other countries, the long term benefits of such steps will increase if they can provide incentives for behavioural change and avoid the moral hazard of excessive coverage.

A remaining question relates to the involvement of the private sector, beyond large corporations, infrastructure operators, and large banking networks. As in many countries, the informal sector, artisans, merchants, transport companies, small industries or subcontractors and small service businesses are poorly prepared and insured to confront a natural or industrial disaster. This question has been partially taken into account in Morocco by the Central Guarantee Fund and through a section of the Fund to Combat the Effects of Natural Disasters.

Concerning the financial contingencies covered by the government and the various ministerial sectors, particularly concerning damage to public infrastructure, an integrated risk management strategy may also include dedicated financing tools and eventually risk transfer mechanisms. Finally, this raises the issue of constructing a dedicated structure for the risk of significant losses to the government's budgetary resources in a major disaster.

How much compensation should remain the responsibility of the general budget compared to an annualised financial package to absorb prevention expenditure and reconstruction costs? Part of the answer is illustrated by the amount the Moroccan government has already spent on recent disasters and the associated one-off expenditure. The MnhPRA model for evaluating damage to public infrastructure may allow for testing various hypotheses to evaluate the appropriateness of coverage mechanisms.

Recommendations

CONSOLIDATE AND STREAMLINE NATURAL DISASTER FINANCING SYSTEMS TO PROMOTE RESILIENCE

Based on financing mechanisms already established in Morocco for recovery and reconstruction (as well as prevention), consolidating existing tools and fine-tuning financial strategies can adjust resources to needs over the medium and long term.

Establish annual financing in the government budget (a minimum fixed percentage, for example) that will eventually smooth prevention and reconstruction financing

There is a close relationship between a country's acceptable level of protection in the medium and long terms (number of deaths avoided, benchmark flood, seismic scale, etc.) and the need for investment in prevention, crisis management and reconstruction. For that reason, an optimal outcome can be achieved by an appropriate preparation and though high-level decisions, following a simulation of needs and of possible financing means.

Use risk assessment to better estimate future financial needs.

Improved advance understanding of risks, damage and their geographic distribution, as recommended for evaluation and prevention (in addition to the risk atlas and based on Desinventar), is a preliminary condition for the estimation of financial need and for the optimal division between various budgetary and other resources.

The MnhPRA model, or its equivalent, may provide significant assistance to assess damage and test various financing hypotheses. To this end, such models must be powered, calibrated and updated with relevant data and comparisons. This also illustrates the importance of data availability and access.

Pool budgetary mechanisms for compensation, using clear rules concerning the use of public funds.

The creation of FLCN in 2009 must be followed by flexibility in emergency priorities, along with transparent rules for prioritising sectors and sites based on cost-benefit analysis for bid requests associated with a decentralisation policy.

FLCN has proved its success and relevance, and its services will increasingly be requested as the fund continues to become more known and appreciated. Beyond the question of resources, the way in which competing projects are negotiated must be explained and justified. Below a certain amount, it may be possible to decentralise these negotiations to the local level. That is why the rules, specifications and evaluation methods must be coordinated and understood.

Consolidate and extend the insurance system for the agricultural sector and accelerate the implementation of coverage mechanisms for individuals.

Proposals to extend the agricultural insurance coverage system (geographically, expanding lists of covered industries and risks) and to improve its financing using a transfer of risk are welcome, given its quality and efficiency.

The insurance and aid mechanisms for individuals already underway must be acknowledged and pursued with determination, while avoiding the effects of moral hazards, as much as possible.

The low penetration rate of insurance for individuals is not favourable for a rapid accomplishment of housing repairs or reconstruction. An insurance mechanism such as that outlined in Law 110-14 could help achieve this objective while alleviating strain on public finance. In addition, this mechanism is a learning factor if it can eventually be supplemented by some *bonus-malus* type incentive mechanisms.

Promote "Build Back Better"

It is crucial the remarkable initiatives for relocation or the rejection of rebuilding on-site, are amplified by an approach such as "Build Back Better" to reduce the future risk of recurrence and to take advantage of the opportunity to rebuild in a more resilient and sustainable manner.

Low interest rates and long maturities that are relatively easy to obtain from national or international financing sources could lead to "better" reconstruction, including in the residential sector: higher costs for construction are balanced by improved resilience, as they build longevity and reduce operating costs.

Bibliography

El Azizi, A. (2010), Tangier – The Black Book of Floods, Issue 31, 23 January 2010, http://www.actuel.ma/index.php?option=com_magazines&view=detail&id=77, reviewed 14 December 2015. Behind the Scenes in Tangier, "Tangier: Flood Project", http://www.lescoulissesdetanger.com/actualites/124-tanger-se-mobilise-contre-les-inondations-pour-mettre-a-labri-son-tissu-industriel.

Sidi Bernoussi Industrial District Economic Operators Association (IZDIHAR) (2010), "The Sidi Bernoussi Industrial District Bandages its Wounds", 12 May 2010, IZDIHAR, http://izdihar.info/uploads/presse/f53.pdf.

El Yacoubi, M. (2014), Risk Management Strategy in the Agricultural Sector, presentation made to the International Conference on Insurance and Reinsurance of Agricultural Risk, 30 January 2014, Marrakesh.

Michel-Kerjan, Erwann O. (2014), CAS-FLCN: Information items for the definition of priorities and selection criteria for projects submitted for financing and operational implementation, report produced as part of the World Bank's assistance to the Ministry of the Interior.

Ministry of Economy and Finance (2012), Draft Finance Law for fiscal year 2013: Report on special treasury accounts, Ministry of Economy and Finance, www.finances.gov.ma/Docs/2012/db/7644_cst_fr.pdf.

Ministry of Economy and Finance (2013), Draft Finance Law for fiscal year 2014: presentation note, Ministry of Economy and Finance, www.finances.gov.ma/Docs/2013/db/np_fr.pdf.

Ministry of Economy and Finance (2014), Draft Finance Law for fiscal year 2015: presentation note, Ministry of Economy and Finance, www.finances.gov.ma/Docs/2014/DB/np_fr.pdf.

Ministry of Economy and Finance (2015a), Note on the presentation of the draft law establishing the coverage system for the consequences of disasters and modifying and supplementing Law no. 17-99 on the insurance code. www.sgg.gov.ma/portals/0/AvantProjet/118/Avp_loi_cat-nat_Fr.pdf, reviewed 18 December 2015.

Ministry of Economy and Finance (2015b), Draft Finance Law for fiscal year 2016: presentation note, Ministry of Economy and Finance, www.finances.gov.ma/Docs/DB/2016/np_fr.pdf.

Ministry of Public Works, Transport and Logistics (2015), 2014 Annual Report: Large construction projects and reforms in the public works, transport and logistics sectors, Ministry of Public Works, Transport and Logistics.

OECD (2012), "Disaster risk assessment and risk financing", *G20/OCDE Methodological Framework on Disaster Risk Assessment and Risk Financing*, OECD Editions, Paris, http://www.oecd.org/gov/risk/G20disasterriskmanagement.pdf.

Sehimi, M. (2014), Construction Sites for a Billion, Maroc Hebdo, 8 May 2014, www.maghress.com/fr/marochebdo/118431, reviewed 19 February 2016.

Zurich (2015), Floods in Morocco in 2014: What Lessons Can Be Learned From Guelmim and Sidi Ifni?, www.zurich.com/_/media/dbe/corporate/docs/corporate-responsibility/risk-nexus-morocco-floods-of-2014-november-2015.pdf?la=en

Appendix A.

Methodology

This report is the product of a "peer review" conducted using the OECD's own principles. It was directed by the OECD Secretariat between December 2014 and May 2016, included three peer experts in risk management in OECD countries, and received support from the Moroccan Ministry of General Affairs and Governance and the Ministry of the Interior. It was financed in part by the Global Initiative on Disaster Risk Management (GIDRM) and the German Agency for International Cooperation (GIZ).

This study is based on an established method for information and data gathering, and on an analytic framework used in previous studies on risk management policies in Italy, Japan and Norway. Its purpose is to identify the strengths of public risk management policies in Morocco and the issues still to be resolved, as well as to determine actions to be taken to improve risk management and, finally, to assess the country's resilience in coping with major risks.

The study's structure was defined in close consultation with the Moroccan Ministry of General Affairs and Governance and the Ministry of the Interior. The analysis was conducted on both a national and regional scale, and included the implementation of policies on the local level. It focused on four high priority risks: floods, droughts, earthquakes and tsunamis. Finally, it took into account the entire risk management cycle (risk assessment, prevention and mitigation, crisis response and management, and recovery and rebuilding) and paid particular attention to questions of governance and inclusion.

This study was built around a broad and inclusive dialogue, with contributions from a wide range of stakeholders in ministerial departments, public institutions and entities, research institutions and universities, the private sector and civil society. A launch event was held in Rabat on 10 April 2015, presenting the study's objectives and the approach used by the OECD Secretariat. This event brought together over 80 participants, and resulted in a rich and fruitful experience. A self-assessment questionnaire was sent to stakeholders for an overview of the risk management structure in Morocco. Eighty-nine responses were received, of which 25 came from public authorities at the central level, one from the regional level, 57 from the local level and six from economic players. Their responses provided initial information to the experts and peers participating in this study, and contributed to the preparation of exploratory missions.

Two exploratory missions, at the central and regional levels, involved OECD experts and peers (from Germany, France and Switzerland) contributing to this study. Following the analysis, these missions allowed hypotheses to be tested against the realities on the ground. The first mission took place between 18 and 22 May 2015, and included meeting with a wide range of key players working at the central level. In total, 33 interviews were conducted during this mission. A second mission took place between 8 and 12 June 2015, to evaluate the implementation of national policies at the local and regional level. Six regions were visited, with distinct risk and vulnerability profiles: Tangier-Tétouan, where flood and earthquake risks are present, Agadir, with earthquake and tsunami risks, Rabat and Casablanca, where a significant portion of the population is concentrated and where risks of floods and tsunamis exist, Haouz province, with flood and drought risks, and Kenitra and the Gharb plain, also facing risks of flooding and drought.

Preliminary results were presented for discussion during the national policy dialogue meeting held 13 October 2015, in Rabat. The event included almost 200 participants, representing various public, private sector and civil society organisations involved in risk management in Morocco. Exchanges with the participants helped to refine the study's preliminary results through participative dialogue, making the recommendations more appropriate to the realities in Morocco.

Based on the information gathered during these stages, the OECD Secretariat conducted the analysis according to the principles set forth in the OECD recommendation on major risk management. The factual portion of the resulting preliminary version of the report was sent to all stakeholders for comments in February-March 2016. Comments from almost 30 players were reviewed and incorporated into the report. The report's results were discussed during a final presentation event in Morocco in May 2016.

Appendix B.

List of organisations encountered

Government agencies and public institutions
Ministry for General Affairs and Governance
Ministry of the Interior
Local Communities Department
Directorate-General of Civil Defence
Monitoring and Coordination Centre
Ministry of Agriculture and Maritime Fishing
Ministry of Economy and Finance
Insurance and Social Contingency Department
Ministry of National Education and Professional Training
Ministry of Energy, Mines, Water and the Environment, Ministry for Water
Hydraulic Infrastructure Department
Water Research and Planning Department
National Meteorological Office
Hydraulic Basin Agencies (Sebou, Tensift)
National Electricity and Potable Water Office
Ministry of Energy, Mines, Water and the Environment, Ministry for the Environment
Monitoring and Risk Prevention Department
Ministry of Public Works, Transport and Logistics
National Ports Authority
National Airports Authority
Morocco Roads
National Railway Authority
Ministry of Housing and Urban Policy
Al Omrane Holdings
Ministry of Industry, Commerce, Investment and the Digital Economy
Ministry of Health
Ministry of Tourism
Ministry of Urban Development and Regional Planning
Urban Agencies (Tétouan, Al Hoceima, Agadir)
High Commission for Waterways and Forests and Fight Against Desertification
Agency for the Promotion and Development of Northern Morocco

Regional bodies
Wilayas (Tangier-Tétouan, Greater Casablanca, Souss-Massa)
Provinces and Prefectures (Al Haouz)
Communities (Tétouan, Rabat, Ourika)
Private sector and insurers
Bank Al Maghrib
Central Guarantee Fund
Moroccan General Business Federation
Lydec
Maroc Telecom
Royal Moroccan Phosphate Office
Redal
Société Centrale de Réassurance
Civil society
Ribat Al Fath Association for Sustainable Development - Rabat
Life and Earth Sciences Association
Moroccan Red Crescent
Research institutions
Economic, Social and Environmental Council
National Centre for Scientific and Technical Research
Royal Centre for Remote Sensing in Space
Economic, Social and Environmental Council
Hassan II Agronomic and Veterinary Institute (Drought Observatory)
Royal Institute for Strategic Studies
Scientific Institute - Mohammed V University Rabat
Ibn Zohr University of Agadir
Sciences Faculty of Rabat

Appendix C.

Selection of identified best practices

The following table presents a non-exhaustive list of best practices for risk management identified in Morocco by the OECD's international team of experts:

List of best practices in risk management in Morocco

List of best practices in risk management in Morocco	
Crisis management coordination	CVC
Integrated approach to forest fire management	HCEFLCD
Simulation of tsunami impacts	HCEFLCD
Flood risk atlas of Haouz Province	ABH Tensift
Prospective approach to risk assessment	IRES
Community guide to prevention against natural risks in Tétouan	Tétouan Municipality
Making children aware of earthquake risk by the Agadir Life Sciences Association	Souss Life and Earth Sciences
Reduction in exposure of vulnerable populations through a relocation programme	Idmaj Sakane
Maintenance of "no-build" zones in Agadir due to seismic risk since the 1960s	Agadir Municipality
Use of satellite imagery for urban management by CRTS contract – Urban Agencies	CRTS/Urban Agencies
Public-private partnership work for the development of paraseismic code RPS 2011	Ministry of Housing
Public-private participative prevention financing for the Super Collector west of Casablanca	Casa Development
Call for prevention projects by the Fund to Combat the Effects of Natural Disasters	Ministry of the Interior
Business continuity plan	Bank Al Maghrib
Network redundancy systems	Maroc Telecom
Multi-agency crisis management coordination	Ministry of Public Works
Early warning system for floods in the Ourika Valley and connection with populations	ABH Tensift/Japanese Cooperation
Direct transmission of weather warnings to critical infrastructure operators	DMN
Emergency response plan for water agencies	LYDEC and ONEE
INSARAG Certification for Civil Defence	DGPC
Development of volunteer emergency services in the Fez and Meknes areas	DGPC/Swiss Cooperation
Reconstruction standards for rail lines	ONCF
Parametric insurance in the agricultural sector	MAPM

Appendix D.

Questionnaire results

The questionnaire presented in Appendix E has led to the development of an initial evaluation of the level of risk knowledge, awareness and preparation among the various stakeholders involved in dealing with risk, particularly ministerial departments, public institutions and entities, research institutions and universities, the private sector and civil society. Questionnaires were sent to stakeholders in April 2015, and 89 responses were received by the OECD Secretariat, including 25 from central government agencies and 58 from the regional level.

At the central level, the main trends drawn from the responses showed the importance given by central players to the challenges of risk management. Eighty-eight per cent of respondents[1] confirmed the legal and institutional structure for risk management is not up to the challenge involved. The questionnaire also helped highlight information on risk assessment: although 83% of the central public agencies have a multi-risk assessment process, 67% confirm the evaluation is not regularly updated. In addition, among the most striking results, 88% of organisations questioned believe that the public is only slightly aware, or not aware of the risks they are subject to, revealing significant gaps in awareness. Regarding prevention policies, only 29% of respondents think that ground use policies are effective. However, 71% think construction standards are effective. As to emergencies, 43% of respondents think the level of communication and coordination between emergency actors is appropriate, resulting from the establishment of the Ministry of the Interior's Centre for Monitoring and Coordination in 2007, responsible for multi-actor emergency coordination.

Different issues emerged from the analysis of the questionnaire responses at the regional level. Regarding evaluation, almost one regional public agency out of two challenges the existence and implementation of risk assessment procedures, revealing true gaps in this area. Evaluations conducted are primarily based on studies, particularly those conducted jointly with the Hydraulic Basin Agency concerning flood risk, or on experience (for example, an area is considered a flood zone if it has flooded in the past). As for prevention, local agencies highlight a lack of resources for preventive structures: 59% confirm they have not included structural policies in their prevention efforts. Non-structural policies are preferred (77% of respondents have non-structural measures in their organisations), which testifies to the importance of the role of local organisations and local public agencies in making the population aware of the natural risks. Concerning emergency response, the vast majority of local organisations questioned (70%) confirm

[1] For reasons of convenience, percentages are used, but these should not be interpreted in a statistical sense, since the sample size of responding institutions is not representative.

they have not developed emergency plans for risks, which demonstrates the ORSEC Plan, established at the national level, does not yet have local parallels.

Appendix E.

Questionnaire sent to stakeholders[1]

1. Legal and institutional structure for risk management in Morocco

Risk management governance: this part describes the risk management system and its principal players as well as the current legislative and regulatory framework. It is focused on clarity in the definition of roles and responsibilities, the organisation's coherence, its ability to achieve established goals and its ability to grow and change. The study examines the effectiveness of procedures for risk management and offers solutions for their implementation through improved governance.

1.1 What are your organisation's roles and responsibilities in decision making and implementation for the management of major risks in Morocco? Please specify at which level *(national, regional, local).*

1.2 Are these roles and responsibilities specified by a national crisis management strategy or other legal or regulatory documents? *If yes, please briefly describe the main points.*

1.3 Has your organisation contributed to the development of public policies related to risk management in Morocco? *If yes, please describe the role it has played.*

1.4 In your opinion, do these documents sufficiently define the goals to be achieved in long-term risk management and the means to achieve them through public policy in general, and for your organisation in particular? Why or why not?

1.5 From your point of view, are the connections between risk management policy and sector policies sufficiently utilised? *Please describe the sector policy in which you are involved (e.g., water management policy, regional development policy, urban development policy, agricultural policy, or others).*

1.6 Is there a structure in place at the national level responsible for coordinating all public policies and stakeholders for crisis management in Morocco to which you contribute? *If yes, please briefly describe it, its role and how it operates.*

1.7 With which other entities does your organisation coordinate to implement its roles and responsibilities (sector ministries, public agencies, local governments and organisations, businesses, associations) in the various areas of risk management? *If you have a responsibility in the areas listed below, briefly describe how this cooperation is structured, including the entities with which you cooperate and the nature of this cooperation.*

- Risk assessment
- Risk reduction and prevention
- Emergency preparation and response
- Recovery and rebuilding
- Other (please describe)

1.8 From your point of view, is your organisation and, more broadly, all stakeholders, sufficiently taken into account in the development and implementation of risk management policies? *If no, please describe which stakeholders are not taken into account.*

1.9 What resources and budgetary means does your organisation have to fulfil its roles and responsibilities in risk management? *Please describe how these resources are allocated.*

Self-assessment

1.10 Is the legal and institutional structure for managing major risks in Morocco at the appropriate level, given the stakes involved?

YES_ NO_

Why or why not?

1.11 What are the principal challenges to improving the governance of risk management in Morocco?

1.12 In your opinion, are there turf issues between the various stakeholders (duplication, lack of cooperation)?

2. Risk assessment

Risk assessment: This part of the study evaluates the recent advances in Morocco related to the evaluation of socio-economic and infrastructure vulnerability, particularly in hazard mapping, as well as the estimation of risk levels, taking into account the experience gained through cooperation with the World Bank. It is also focused on accounting for future and prospective developments and the use of risk assessment tools to inform the various areas of risk management policy. It includes an evaluation of the contributions of Moroccan scientific research in this area.

Roles and responsibilities

2.1 What are your organisations' roles and responsibilities in evaluating the following risks?

- Flood risk

- Earthquake risk

- Drought risk

- Tsunami risk

- Other natural disaster risks

2.2 Do these roles and responsibilities contribute to an integrated evaluation process for all risks at the national level? *If yes, please describe the main areas involved.*

and answer the following questions:

- Is the risk assessment a multi-risk process? YES_ NO_
- Is the risk assessment updated regularly?YES_ NO_

2.3 Which players does your organisation regularly work with on risk assessment? *Please specify the coordination and cooperation mechanisms for each of the following categories.*

- National institutions

- Local institutions

- Businesses

- Associations

2.4 Are these coordination mechanisms sufficiently effective for evaluating all major risks in Morocco by using the best available resources? *Please specify areas for improvement.*

Data for risk assessment

2.4 Do you have systems in place for measuring, archiving and sharing data on hazards? *If yes, please describe the main approaches for the following risks:*

- Flood risk - Earthquake risk - Drought risk - Tsunami risk - Other natural disaster risks

2.5 In your opinion, are there areas for improvement in the ability to measure, archive or share data related to hazards? *If yes, please specify.*

2.6 Is information related to the human, economic, environmental or cultural stakes at serious risk available for risk assessment? *If yes, please specify which are used and in what form.*

2.7 Does risk assessment include the following specific vulnerabilities? *If yes, please describe how.*

- Vulnerable populations - State of buildings - State of infrastructure

2.8 Are damages and losses from disasters systematically collected in a data base? *If yes, please specify the categories of losses and damages used (lives, property, infrastructure, environment, etc.) and the methods used for information gathering.*

2.9 Do you include prospective items for longer term risk assessment (climate change impact, demographic trends, urban expansion)?

Risk assessment tools and methods

2.10 What risk assessment methods are used by your organisation?

Please indicate whether they include the following items and describe.

- Risk assessment in terms of impacts and probability

- Evaluation of secondary risks linked to an interruption in activity (e.g., loss of business, network interruptions, impacts on value chains, etc.)

2.11 Are the results of risk assessments aggregated into a national risk assessment that includes comparisons among major risks in terms of impact and probability? *If yes, please describe how.*

2.12 Has your risk assessment included developing a risk map? *If yes, please describe the types of maps developed and whether they meet a regulatory requirement.*

Use of risk assessment in decision making

2.13 Are risk assessment results used in the following actions? *If yes, please describe how for each item.*

- Risk awareness among the public and stakeholders

- Prioritising prevention activities

- Developing emergency plans

- The budget process and resource allocation

- Developing financial contingency mechanisms

Self-assessment

2.14 In your opinion, what are the main challenges in evaluating major risks in Morocco?

3. Prevention policies

Prevention policies: In this section, the study looks at the effectiveness of various structural and non-structural measures for reducing risk exposure and vulnerability of populations, critical infrastructure and businesses, the agricultural sector and public institutions. A specific focus is on critical infrastructure and the private sector.

Communication and awareness of risks

3.1 What are your organisation's roles and responsibilities in improving awareness and understanding of risks by stakeholders (individuals, businesses, local communities)?

3.2 Are these roles and responsibilities specified in one or more regulations on communication of risks? *If yes, please specify which regulations.*

3.3 Which other actors play an important role in communicating and raising awareness of major risks? How is coordination between the players ensured?

3.4 To the best of your knowledge, do measures related to communication, information and risk awareness in Morocco include the following provisions? *If yes, please specify which measures and describe the provisions.*

- Specific measures vis-à-vis vulnerable groups
- Actions directed towards businesses and economic players
- Information not only about risks but also about measures to be taken in emergencies or for risk prevention
- Regular information campaigns
- Use of multiple means of communication, including new technologies
- Evaluation measures for the effectiveness of risk awareness programmes

Self-assessment

3.5 In your opinion, are populations living in risk areas sufficiently aware of the risks and informed about measures to take in the event of a disaster and measures available to reduce risk?

Very aware ☐

Aware ☐

Not very aware ☐

Not aware ☐

3.6 In your opinion, what can be done to improve awareness of major risks in Morocco?

- Among the public - Among businesses - Among local authorities

Structural prevention measures

3.7 What are your organisation's roles and responsibilities in the design and implementation of structural measures for reducing major risks in Morocco, such as protective infrastructure (floods, droughts, storms, tsunami)?

3.8. Which other players have a significant role in the development of structural prevention measures for major risks in Morocco? How is coordination among the players ensured?

3.9 Do you have information on financial resources allocated for the development of structural protection measures for major risks? *If yes, please indicate the annual amounts, financing sources and changes over recent years.*

3.10 What are the decision making mechanisms for resource allocation in the development of protective infrastructure for major risks? *Please describe the decision making process and specify the tools used to support decision making (cost-benefit studies, multi-criteria analyses).*

3.11 Do investment decisions for structural measures take into account the possible consequences of climate change? *If yes, please describe how.*

3.12 Does your organisation contribute to the maintenance of protective infrastructure? How is this monitored?

Self-assessment

3.13 In your opinion, what are the main challenges in the implementation and maintenance of structural prevention measures for major risks in Morocco?

3.14 In your opinion, are structural prevention policies for major risks in Morocco at the appropriate level, given the stakes involved?

YES_ NO_

Please elaborate on your answer.

Non-structural prevention measures

Zoning and ground use

3.15 What are your organisation's roles and responsibilities in the design and implementation of ground use and urban development policies related to improved resilience to major risks?

3.16 Which other organisations play an important role in the design and development of ground use and urban development policies? How is coordination between different players ensured?

3.17 What are the criteria for ground use related to the major identified risks in Morocco? *If possible, please list the associated regulatory documents.*

3.18 Are there specific criteria related to critical infrastructure (water, electricity, transport, telecommunications, security)? *If yes, please specify.*

3.19 What resources are available for monitoring the implementation of ground use and urban development policies?

3.20 In your opinion, are ground use and urban development policies for major risks sufficiently effective to improve resilience?

YES_ NO_

Please elaborate on your answer.

Construction codes and standards

3.21 What are your organisation's roles and responsibilities in the development and implementation of construction codes and standards for improving resilience to major risks (particularly for earthquakes)?

3.22 Which other organisations play an important role in the design and development of construction codes and standards? How is coordination between different players ensured?

3.23 What criteria are used for construction codes and standards for the major identified risks in Morocco? Are there specific criteria for critical infrastructure?

3.24 What resources are available for monitoring the effective implementation of construction codes and standards?

3.25 In your opinion, are construction codes and policies sufficiently effective to improve resilience to major risks?

YES_ NO_

Please elaborate on your answer.

Self-assessment

3.26 In your opinion, what improvements could be made to ground use and urban development policies and construction codes and standards to reduce vulnerability to major risks in Morocco?

Early warning systems

3.27 What are your organisation's roles and responsibilities in the implementation of early warning systems in Morocco for major risks?

3.28 Which other organisations play an important role in early warning systems and how is coordination between different players ensured?

3.29 Do you have systems in place to provide rapid detection of all major hazards? *If yes, please describe them.*

3.30 Do you have communication channels for disseminating warnings to threatened communities? *If yes, please describe them.*

3.31 To what extent do warnings lead to the activation of preparation levels and response plans by emergency services?

Self-assessment

3.32 How effective do you think current early warning systems are? *Please elaborate on your answer.*

4. Preparation for emergency management

Preparation for emergency management: The study is focused on the ability of the response system to account for new complexities in major crises and in their consequences. Coordination between the various players in the multi-agency response system and their joint ability to interact to improve response will be evaluated. The use of new technologies and investments in preparation, training and drills, the ability to cooperate with the private sector, volunteer organisations and international aid groups will be analysed. In particular, the coordination and communication system used during crisis management and the ORSEC response plan evaluation will be assessed.

4.1 What are your organisation's roles and responsibilities in emergency preparation and response?

4.2 How is your organisation structured to fulfil these roles and responsibilities and what resources are allocated for that purpose?

4.3 With which other players in the national government, regional bodies and businesses does your organisation cooperate in emergency preparation and response, and how is that cooperation structured?

4.4 How would you characterise the level of coordination and communication between actors involved in emergency preparation and response?

SUFFICIENT_ INSUFFICIENT_

Please describe why.

4.5 In your opinion, do emergency plans in place in Morocco have the following characteristics:

- Do they cover all major risks?
- Do they include specific vulnerabilities (vulnerable groups, critical infrastructure, schools, hospitals, etc.)?
- Are they frequently tested during drills?
- Do they provide for good cooperation among responding organisations?
- Do they allow for ramp-up based on subsidiarity?

4.6 To what extent do businesses and the private sector participate in emergency preparation and response?

4.7 To what extent are volunteer organisations involved in emergency preparation and response?

4.8 Do emergency response measures plan specific procedures for international cooperation? *If yes, please specify.*

Self-assessment

4.9 How do you rate your organisation's ability to contribute to emergency response to major risks in Morocco?

4.10 What is your evaluation of the current overall capacity for emergency management of major risks in Morocco?

5. Recovery and rebuilding

Recovery and rebuilding: In this portion, the study is focused on continuity of activities for more rapid recovery, particularly for businesses and public services. Public compensation and aid mechanisms will be analysed, as will the possible role of insurance in various sectors, particularly in agriculture.

Continuity of activity

5.1 Does your organisation encourage the development of plans for continuity of activity in businesses and public services?

5.2 Which public policy tools are in place to encourage businesses to develop plans for continuity of their activity in dealing with major risks?

5.3 Does your organisation discuss emergency plans with critical infrastructure system operators and other economic players? *If yes, please describe how this consultation takes place.*

Compensation and insurance

5.4 What are your organisation's roles and responsibilities relative to compensation for damage resulting from disasters to:

- Individual households
- Non-agricultural businesses
- Agricultural businesses

- Public infrastructure

5.5 How do these compensation mechanisms function?

5.6 Are there other insurance or financial mechanisms in place to compensate for economic losses caused by major risks? *If yes, please list.*

5.7 To your knowledge, what is the penetration rate for major risk insurance among individuals and businesses?

5.8 Are there public policies for increasing the rate of insurance coverage for disaster risks among individuals or businesses?

Self-assessment

5.9 In your opinion, is the current system for disaster loss and damage compensation adequate to deal with major risks in Morocco? *Please elaborate on your answer.*

Feedback and organisational change

5.10 What are your organisation's roles and responsibilities related to feedback following major disasters?

5.11 Are you aware of examples in which feedback has led to organisational or legislative change in the risk management system in Morocco? *If yes, please describe.*

Note

1 Four versions of this questionnaire were prepared based on the respondent's role: central public authorities, regional authorities, local authorities, economic players. The questionnaire presented in this appendix was sent to central authorities.

Appendix F.

Description of principal laws and regulations

Legal text	Main provisions
Dahir dated 30 April 1955 related to civil defence	The 30 April 1955 Dahir entrusted Moroccan Civil Defence with providing emergency services in disasters : "the goal of civil defence, in time of peace, is to implement and coordinate emergency services in major disasters." (Article 1).
Decree no. 2-97-176 dated 15 December 1997, relative to the authority and organisation of the Ministry of the Interior *Supplemented by decree no. 2-04-750 dated 27 December 2005, decree no. 2-05-1585 dated 28 June 2007 and decree no. 2-08-159 dated 16 January 2009*	Decree no. 2-97-176 defines the authority of the Ministry of the Interior, responsible for the domestic administration of the Kingdom, ensuring public order, informing the government and ensuring the supervision of local bodies (Article 1). The DGPC is defined as a body for reflection, research, study, consultation and intervention for the defence of persons and their property in all circumstances (Article 36).
Minister of the Interior Order no. 1502-02 dated 23 October 2002, relative to the creation, organisation and authority of services outside the Directorate-General of Civil Defence and their regional responsibilities	Order no. 1502-02 defines the organisation of civil defence at the regional level : external services are created, formed from regional and provincial commands (Article 2). They are responsible, in their assigned regions, for organising emergency services and for taking all preventive measures to make their regions less vulnerable to risks.
Dahir including Law no. 1-75-168 dated 15 February 1977, on the authority of governors	Dahir including Law no. 1-75-168 defines the authority of governors. The governor ensures the application of Royal Decrees, laws and regulations and the execution of government decisions and directives in the prefecture or province (Article 2). He is responsible for maintaining order (Article 3). He coordinates the external activities of government agencies, public establishments whose activity area does not go beyond the prefecture or province (Article 5).
Organic Law no. 111-14 relative to regions	Organic Law no. 111-14 issues all regulations related to the organisation of regions. The regions are responsible for the promotion of integrated and sustainable development, through the best use, exploitation and preservation of natural resources (Article 80). They are responsible for the development and follow-up on the execution of the regional development programme and the regional development plan – reference document on development (Article 81). Also, among its shared responsibilities, the region is responsible for flood prevention, preservation of protected areas, preservation of forest ecosystems and preservation of water resources (Article 91). Among its delegated responsibilities are the regional management of facilities and infrastructure, especially those related to commerce, health, water and the environment. (Article 94).
Organic Law no. 112-14 relative to prefectures and provinces	Organic Law no. 112-14 issues all regulations related to the organisation of prefectures and provinces. These are responsible, in their own jurisdictions, for promoting social development, particularly for making available basic facilities and services and fighting poverty (Article 78). They participate in raising the level of rural health care, training and infrastructure (Article 86). They are responsible for constructing and maintaining small and medium-sized water projects, particularly in rural settings (Article 89).

Organic Law no. 113-14 relative to municipalities	Organic Law no. 113-14 issues all regulations related to the organisation of municipalities. These are responsible for the organisation, coordination and follow-up for local services (Article 77). They implement the community's Action Plan (PAC), based on a participative approach, establishing planned development activities for the next 6 years, including a needs and opportunities assessment for the municipality (Article 78). Municipalities create and manage community services and infrastructure, particularly for water and electricity, urban public transport, liquid and solid waste management, public hygiene, and in emergencies, transporting the injured (Article 83). In addition, the municipality monitors compliance with regulations, such as all urban and community development documents (Article 85). Among the responsibilities shared with the national government, municipalities take part in protecting the environment, shoreline management, development of parks, cliffs, lakes and river banks, maintaining national roads and developing medinas (Article 87). Finally, they are responsible for the construction and maintenance of small and medium sized water works within their jurisdiction (Article 90).
Decree no. 2.13.253 dated 20 June 2013 et Decree no. 2.13.836 dated 13 November 2013	These decrees establish the authority of the Minister attached to the Head of Government responsible for General Affairs and Governance: he is responsible for defining a national strategy for coping with risks, with the exception of those related to defence and security, and for encouraging cooperation with government institutions and organisations.
Decree no. 2-94-830 dated 20 January 1995 establishing the organisation and authority of the Ministry of Housing	This Decree defines the authority of the Ministry of Housing and Urban Policy. It is responsible for defining strategies for intervention and encouragement for construction and real estate development and monitoring their completion. The technical housing department's mission is standardising and promoting quality and innovative building techniques and procedures, as well as their dissemination. The Ministry is also responsible for developing and applying regulations on construction safety and safety policy (Article 11).
Decree no. 2-04-267 dated 10 May 2004 modifying Decree no. 2-02-177 approving paraseismic construction regulations (RPS 2000) applicable to buildings, setting paraseismic rules and establishing the national Paraseismic Engineering Committee	Decree no. 2-04-267 establishes the mandatory RPS 2000 paraseismic construction regulation for the entire country. This regulation describes design criteria and technical procedures to limit building vulnerability to earthquakes. Its scope includes reinforced concrete and steel structures in current use, with simple and regular configurations.
Circular from the Minister for Housing and Urban Development no. 824/ 2173 dated 20 February 2003 relative to "the creation of the regional protection plan for floods and the provincial or prefectorial water commission"	Circular 824/ 2173 coordinates actions of various local players related to flood prevention. It includes the creation of provincial and prefectorial commissions for flood prevention, which bring together the appropriate parties (hydraulic basin agencies, urban agencies, regional housing authorities and development agencies, national meteorological, public works and transport agencies, civil defence and Royal Gendarmerie, etc.)
Law no. 10-95 on water *Established by Dahir no. 1-95-154 dated 16 August 1995 as supplemented by Law no. 42-09, established by Royal Decree no. 1-10-104 dated 16 July 2010*	The Water Law governs national water policy based on a prospective vision and taking into account changes in resources as well as in national water requirements. It includes legal provisions for streamlining water use, broadening access to water, inter-regional solidarity with the purpose of ensuring water security across the Kingdom. It provides specific measures for flood and drought risks. It regulates the implementation of flood control measures.
Law no. 11-03 relative to the protection and development of the environment *Established by Dahir no. 1-03-59 dated 12 May 2003*	Law no. 11-03 issues rules for national policy on the protection and development of the environment, with the goal of protecting the environment from all forms of degradation, improving the public's living standards and situation and establishing a system of responsibilities providing for repair of damage to the environment and compensation for victims. Law no. 11-03 provides for mechanisms to manage and protect the environment : specific protection measures are ordered to counter desertification, floods, the disappearance of forests, erosion, loss of arable land and pollution of soils and other resources. The developer of any classified installation subject to authorisation is required to draw up an appropriate emergency plan including informing competent authorities, neighbouring populations and the evacuation of employees (Article 53).

Law no. 12-03 relative to environmental impact studies *Established by Dahir no. 1-03-60 dated 12 May 2003*	The goal of Law no. 12-03 is to evaluate, methodically and beforehand, possible repercussions of all planned activities, construction projects, development and works, undertaken by any individual or organisation on the environment. The direct and indirect effects are evaluated, particularly on humans, animals, plants, soils, water, air, the climate, the natural setting as well as on public health and safety (Article 5).
Framework law no. 99-12 on the national environmental and sustainable development charter *Established by Dahir no. 1-14-09 dated 6 March 2014*	The National Environmental and Sustainable Development Charter defines the rights and responsibilities of individuals and organisations regarding the protection and development of the environment, as well as the principles and values that should surround public and private policies for sustainable development. It confirms that the country is exposed to natural environmental risks, including those resulting from climate change. Everyone also has the responsibility to protect and improve the environment, to ensure the country's natural and cultural heritage, improve health and the quality of life
Law no. 12-90 relative to urban development *Established by Dahir no. 1-92-31 dated 17 June 1992*	Law no. 25-90 legally defines the various urban planning documents (SDAU, PZ, PA, compliance orders, construction permits) and regulates construction in urban municipalities, within rural community limits, peripheral areas of urban communities and urban development groupings.
Law no. 25-90 relative to housing developments, housing groups and parcels *Established by Dahir no. 1-92-7 dated 17 June 1992*	Law no. 25-90 supplements Law no. 12-90 by legally specifying housing developments, parcels and housing groups, and defining the rights and responsibilities of developers, with criminal penalties.
Dahir including Law no. 1-93-51 dated 10 September 1993 establishing urban agencies	Law no. 1-93-51 defines the role of urban agencies. In particular, these agencies are responsible for conducting the necessary studies for the establishment of urban development master plans and monitoring their execution; preparing draft urban regulatory documents, particularly zoning plans and improvement and development plans; providing an opinion on all urban development projects; monitoring conformance of housing developments, parcels, housing groups and current construction projects with legislative and regulatory requirements and with approvals granted; promoting building renovation and rehabilitation in neighbourhoods lacking infrastructure; and providing technical assistance to local governments on urban planning and development as well as to public and private organisations.
Draft Law establishing the system of coverage for the consequences of catastrophic events, modifying and supplementing Law no. 17-99 including the insurance code	The purpose of this Draft Law is to guarantee a minimum right to compensation for bodily harm or loss of the use of a principal residence in case of a catastrophic event, and to establish an offer for coverage of catastrophic events that may affect persons holding an insurance policy. It provides for a blended system of indemnifying victims of catastrophic events, combining an insurance system for persons with insurance policies and an allocation system for uninsured persons, through the establishment of the "Aid Fund for Catastrophic Events".
Decree no. 2-94-285 dated 21 November 1994 relative to the authority and organisation of the Ministry of Public Health	Decree no. 2-94-285 defines the authority of the Ministry of Health. It is responsible, in case of disaster, to contribute to the development of logistical resources for coping with and participating in providing medical assistance to victims.
Dahir no. 1-58-376 dated 15 November 1958 regulating the right of association	Dahir no. 1-58-376 allows groups of persons to freely associate (Article 2). Regularly established associations may acquire, possess and administer public subsidies, member rights and dues, private sector assistance or international cooperation, and facilities necessary for their activities.
Law no. 15-02 relative to ports and including creation of the National Ports Agency and the Ports Operating Company *Established by Dahir no. 1-05-146 dated 23 November 2005*	Law no. 15-02 establishes the legal status of ports and creates two organisations: the National Ports Agency (ANP) (port authority) and the Ports Operating Company (commercial missions). The Law delineates the ports. The ANP also establishes an internal development plan and operating regulations for each port. Specifically, these define the necessary measures for the application of legislative and regulatory requirements for safety, health, security and environmental protection within the port.

Law no. 54-05 relative to the delegated management of public services *Established by Dahir no. 1-06-15 dated 14 February 2006*	Law no. 54-05 applies to delegated management contracts for public services and works by local governments or groups of governments and public institutions (Article 1). The delegate, a public or private legal entity, assumes the responsibility for the public service in accordance with the principles of user equality, continuity of service and adaptation of technological, economic and social changes. It provides its services at the lowest cost and under the best safety, quality and environmental protection conditions (Article 3). The delegate has rights and obligations, including holding regularly renewed insurance policies covering its civil responsibilities and risks that may result from its activities.
Decree no. 2.02.376 dated 17 July 2002 defining the special status of public educational and instructional institutions	Decree no. 2.02.376 requires that risk be included as part of the instructional programme and that viability and security of the educational infrastructure be considered.

Appendix G.

Advisory bodies and commissions

Risk management body & commission	Established by:	Composition:	Missions:
Monitoring and Coordination Centre	-	Monitoring mode: teams conduct mitigation and preparation activities Crisis mode: Crisis room (sections: operations, planning, logistics and finance, communication)	Direct emergency response on the ground, ensure that resources are allocated (financial and material), and coordinate players.
National and regional environmental impact study committees	Law no. 12-03 relative to environmental impact studies & Decree no. 2-04-563 relative to the authority and function of the national and regional environmental impact study committees	Governmental authorities responsible for the environment, interior, public works and transport, development, urban development, tourism, energy and mines, water, health, agriculture and maritime fishing, industry, justice and HCEFLD	Review environmental impact studies and related documents Provide an opinion on the project's environmental acceptability Participate in developing directives prepared by the governmental authority responsible for the environment and issues related to environmental impact studies
National Paraseismic Engineering Committee	Law no. 12-90 relative to urban policy & Decree no. 2-02-177 dated 22 February 2002 approving paraseismic construction regulations (R.P.S 2000) applicable to buildings, establishing paraseismic rules and establishing the National Paraseismic Engineering Committee	Governmental authorities responsible for urban policy, interior, public works, mines and scientific research University departments, scientific and technical institutes, institutions of higher education and training and concerned professional organisations	Provide an opinion on construction classification and zoning maps in seismic areas Study modifications and suggest improvements to paraseismic construction regulations (RPS2000), taking into account the evolution of knowledge on seismic and geotechnical phenomena as well as paraseismic engineering techniques

High Council for Regional Development	Decree no. 2-01-2331 dated 13 December 2001 creating the High Council for Regional Development	Governmental authorities responsible for development and urban policy, interior, economy and finance, agriculture and maritime fishing, public works, environment, energy and mines, etc. Presidents of regional councils Representatives of public organisations, such as ONEE, ONDA, ANP and ONCF Universities, professional associations, professional organisations or those specialising in development and institutions working in development and the environment	Propose overall direction for development and sustainable development Provide an opinion on plans and national and regional documents regarding development that are submitted to it by the government Monitor the alignment of various options and large projects and their conformance with the principles and direction of development policy
National Housing Council	Decree no. 2-01-1011 dated 4 June 2002 creating the National Housing Council	Governmental authorities responsible for development and urban policy, housing, finance, interior, etc. Public and semi-public organisations, such as the Deposit and Management Fund, the National Social Security Fund, the ONEE, the Community Public Works Fund Financial and professional institutions, such as banks and insurance and reinsurance companies, National Order of Architects, National Order of Surveyors	Define housing direction and strategy Propose measures for promoting housing, particularly social
National and regional environmental councils (CNE)	Decree no. 2-93-1011 dated 20 January 1995 relative to the reorganisation of institutions responsible for the protection and improvement of the environment	Governmental authorities responsible for foreign affairs, interior, health, finance, agriculture and fishing, transport, housing, environment, energy and mines, etc. The National Environmental Council may include representatives of professional associations, private organisations, associations specialising in the environment and sustainable development or scientific institutions as consultants	Ensure the inclusion of environmental concerns in economic and social development processes to assist in achieving sustainable development goals Preserve the ecological balance of the natural environment, particularly water, soil, air, animals, plants and landscape Prevent, combat and reduce all types of pollution and nuisances
Higher Council for Water and Climate	Law no. 10-95 pertaining to water	Government, basin agencies, ONEE, regional agricultural development offices Water users elected by their peers Prefectorial or provincial assemblies Higher education and scientific research institutions	Develop overall direction for national water and climate policy Provide an opinion on national strategy for improving climate knowledge and dealing with climate impacts on the development of water resources, on the national water plan and on development plans including hydraulic basin water resources

Appendix H.

Development of a structure for expenditure related to risk in Australia, Japan and Switzerland

In Morocco, as in many OECD countries, prevention expenditure is often bundled into very diverse programmes, covering other public policy areas and managed by many different players from central administrative agencies to regional associations and even including the private sector. Some pioneering countries, such as Australia, Japan and Switzerland, have implemented a consolidated structure for managing all expenditure related to risk management.

Australia

In 2001, the Council of Australian Governments (COAG), in cooperation with the Australian Regional Transport and Economy Office, collected and analysed data on government expenditure related to risk management at the state/territories and Commonwealth levels. Government expenditure related to risk management was divided according to the phase of the risk management cycle being considered: (1) preparation and response (the idea of creating a category of expenditure that is both pre- and post-disaster was to account for the entire budget allocated to natural disasters), (ii) recovery, (iii) risk mitigation (structural and non-structural prevention programmes), (iv) other natural risk management expenditure (research, administrative functions). Australia chose to account only for natural risks (including forest fires, earthquakes, floods, storms, cyclones, landslides, tsunamis, tornadoes and meteorites). To collect this data, it distributed questionnaires to agencies involved and conducted follow-up interviews. The data collected can be disaggregated based on the level of government and the agency involved.

Japan

Since 2014 and its White Paper on Disaster Management, the Japanese government has established a framework to allow it to assess the evolution in its expenditure associated with disasters from 1962 to 2014. Japan chose to divide its management expenditure into four categories: (i) scientific and technological research, (ii) disaster prevention, (iii) disaster management and (iv) recovery.

Switzerland

In 2007, the "Natural Dangers" National Platform (PLANAT) conducted an inquiry into expenditure related to risk management at the national and regional levels. This inquiry covered natural risks (floods, avalanches, landslides, earthquakes, storms and extreme temperatures). The estimates were based on freely accessible databases on government expenditure and almost 80 interviews. Using these inquiries and interviews, Switzerland was able to integrate expenditure linked to risk management made by private sector players, including businesses, households, and public-private critical infrastructure operators, into its budgetary framework. PLANAT chose to categorise expenditure by institution (central government, federal governments, municipalities and private sector), by hazard and by function (prevention, response, recovery, evaluation and research). However, PLANAT notes that some estimates may be imprecise, with variations of from -5 to +10% for national expenditure, from -10 to +20% for regional expenditure and from -15 to +30% for businesses.

Appendix I.

Framework for risk-related expenditure developed by the OECD

A standardised framework for expenditure related to risk is presented below. It includes data collection based on the phase in the risk management cycle and separates expenditure by pre- and post-disaster, according to government level and hazard.

Categories for prevention based on the OECD budget framework for risk management expenditure

Sub-category of the risk management cycle[1]	By			Hazard Type	Expense Category - Personnel - Administrative - General expenses - Capital investment - Operations & maintenance - Other
	Ministry/Department	*Local authorities[2]*	*Private sector (businesses, individuals)*		
I.1 Strategic planning[3]					
I.2 Identification and assessment of hazard[4]					
I.3 Mapping of risk/hazard					
I.4 Development[5]					
I.5 Planning, development and construction of protection infrastructure					
I.6 Existing protection measures for buildings (houses, etc.)					
I.7 Protection measures for critical infrastructure (energy, water, transport, road networks, NTIC, etc.)					
I.8 Risk awareness raising and communication activities					
I.9 Risk transfers by the private sector					

Notes: 1 Includes "dedicated" expenses (directly linked to risk management) and "included" expenses (not clearly dedicated to risk management, but contributing wholly or in part to risk management); 2 Local level specific to the country (e.g., municipality, department, region, etc.); 3 Includes activities linked to creating risk management strategies, investments or budgetary plans; 4 Includes vulnerability assessments, expenditure for research, development and training in each of the sub-categories; 5 Includes expenses linked to integrating risk management into development, such as creating development plans, changes in regulations and expenses related to their application; 6 Excludes insurance payments by the private sector and households, includes sovereign risk transfer instruments (e.g. purchases of insurance and reinsurance, interest paid for "catbonds") as well as administrative, research and development, training and other expenses for managing public risk transfer operations.

Source: OECD, unpublished, *Improving the Evidence Base on the Costs of Disasters to Inform Better Policy Making for Disaster Risk Management: Toward a Framework for Accounting National Risk Management Expenditure and Losses of Disasters*.

Appendix J.

Glossary

Hazard: a dangerous phenomenon, natural or artificial, which may cause damage or loss to persons, property, infrastructure or the environment.

Risk: the combination of an event's probability and the resulting damage or loss. Damages result from the interaction between the hazard, the exposure and the vulnerability of property, persons, infrastructure and the environment.

Vulnerability: the propensity of persons, property, infrastructure and/or the environment to be affected or impacted by a disruptive natural hazard.

Risk assessment: method for determining the nature and extent of risk through analysing potential hazards and evaluating vulnerabilities that, taken together, may cause damage or loss to persons, property, infrastructure or the environment.

Prevention: the set of actions taken to reduce risk, particularly by protecting from or mitigating damages or loss caused by the combination of a risk and vulnerability. Structural and non-structural measures are considered separately.

Structural measures: engineering or civil engineering measures to reduce exposure to risk by protecting property or communities, or manage the variability of natural phenomena (e.g., dams or levees against floods or high water, protection against rock slides, anti-avalanche barriers).

Non-structural measures: measures designed to reduce exposure to risk and vulnerability through long term planning and adaptation in confronting hazards (e.g., public awareness raising, preparation for emergency situations and early warning systems, regulations for ground use or urban development, construction codes or restoration of ecosystem functions to mitigate extreme phenomena).

Preparation for emergency management: all measures implemented to respond to an emergency situation caused by a hazard.

Recovery: process for the restoration, reconstruction and improvement of systems (including both the population and the environment) affected by a hazard.

Compensation: all financial tools designed to reduce damages and losses incurred as a result of a hazard.

Continuity of activity plan: the strategy and all planned steps to ensure that an organisation can restart and continue its activities following a hazard.

Resilience: the ability to resist and/or absorb a shock, recover normal function or successfully adapt, when faced with an adverse phenomenon or a change in situation.

ORGANISATION FOR ECONOMIC CO-OPERATION AND DEVELOPMENT

The OECD is a unique forum where governments work together to address the economic, social and environmental challenges of globalisation. The OECD is also at the forefront of efforts to understand and to help governments respond to new developments and concerns, such as corporate governance, the information economy and the challenges of an ageing population. The Organisation provides a setting where governments can compare policy experiences, seek answers to common problems, identify good practice and work to co-ordinate domestic and international policies.

The OECD member countries are: Australia, Austria, Belgium, Canada, Chile, the Czech Republic, Denmark, Estonia, Finland, France, Germany, Greece, Hungary, Iceland, Ireland, Israel, Italy, Japan, Korea, Latvia, Luxembourg, Mexico, the Netherlands, New Zealand, Norway, Poland, Portugal, the Slovak Republic, Slovenia, Spain, Sweden, Switzerland, Turkey, the United Kingdom and the United States. The European Union takes part in the work of the OECD.

OECD Publishing disseminates widely the results of the Organisation's statistics gathering and research on economic, social and environmental issues, as well as the conventions, guidelines and standards agreed by its members.

OECD PUBLISHING, 2, rue André-Pascal, 75775 PARIS CEDEX 16
(42 2016 44 1 P) ISBN 978-92-64-27638-3 – 2017

www.ingramcontent.com/pod-product-compliance
Lightning Source LLC
Chambersburg PA
CBHW080536250926
51846CB00016B/1163
* 9 7 8 9 2 6 4 2 7 6 3 8 3 *